A WORLD OF
ANTIQUE PHONOGRAPHS

TIMOTHY C. FABRIZIO AND GEORGE F. PAUL
WITH PHOTOGRAPHS BY THE AUTHORS

Page 1 Photo:
0-1. The "Victrola" (that is, Victor's innovative "internal-horn" talking machine of 1906) was introduced in Great Britain by Victor affiliate the Gramophone and Typewriter, Limited, in 1907, under the designation "Gramophone Grand." The cabinets of the full-sized "Grand" models were rather wide (see figure 3-88), but abbreviated interpretations ("Library Grand," "Bijou Grand") were envisioned from the first. These lesser "Grands" were narrow and deep. From about 1909 comes this member of the "Grand" family, unusual in "Neo-Classical" style. Domestic designs of the mid-to-late eighteenth century were influenced by the evolving science of archeology (though "science" at the time meant hacking your way into ancient sites to locate tidbits that amused you). Decorative elements were borrowed from the ancients — in this particular instance, suggestions of Roman wall paintings, such as those discovered at Pompeii (transposed to the eighteenth century, replicated in the twentieth). The inlay is typical of Georgian and Sheraton design. The veneer is mahogany, and the internal horn is ebonized, another late-eighteenth/early-nineteenth century conceit. Although all "Grand" models were ostensibly associated with a "historical period," some appear to have been available only by special order. As with all talking machine cabinets which purported to represent a design epoch, considerable liberties were customary. Cabinets in the "Grand" line were constructed locally, but the mechanical works were imported from Victor's Camden, New Jersey, plant. The motor, turntable, arm, etc., are the type used in Victor's "VTLA" models, though the hardware finish is nickel rather than gold. A celluloid plate identifies the seller as Selfridge's department store, a significant Gramophone dealer. Selfridge's, founded by an American entrepreneur, opened in London on March 15, 1909. The impact of this enterprise was immediate and considerable — in 1909, after Blériot's record-breaking cross-Channel flight, his monoplane was exhibited at Selfridge's, where it drew 12,000 visitors. Perhaps one of the curious wandered into the Gramophone rooms, and ended up ordering an instrument in "Neo-Classical" style. (Value code: VR)

Copyright © 2007 by Timothy C. Fabrizio &
George F. Paul
Library of Congress Control Number: 2007924218

Designed by John P. Cheek
Cover deisgn by Bruce Waters
Type set in Zurich BT

ISBN: 978-0-7643-2696-7
Printed in China

Published by Schiffer Publishing Ltd.
4880 Lower Valley Road
Atglen, PA 19310
Phone: (610) 593-1777; Fax: (610) 593-2002
E-mail: Info@schifferbooks.com

For the largest selection of fine reference books on this and related subjects, please visit our web site at **www.schifferbooks.com**
We are always looking for people to write books on new and related subjects. If you have an idea for a book please contact us at the above address.

This book may be purchased from the publisher.
Include $3.95 for shipping.
Please try your bookstore first.
You may write for a free catalog.

In Europe, Schiffer books are distributed by
Bushwood Books
6 Marksbury Ave.
Kew Gardens
Surrey TW9 4JF England
Phone: 44 (0) 20 8392-8585;
Fax: 44 (0) 20 8392-9876
E-mail: info@bushwoodbooks.co.uk
Website: www.bushwoodbooks.co.uk
Free postage in the U.K., Europe; air mail at cost.

Contents

Dedication

We dedicate this book to devoted researchers,
single-minded in their exploration of the history of recorded sound, just as we have dedicated the past ten years of our lives to a series of books of which this is the latest installment.

Acknowledgments

Thank you, of course, to all the collectors who have allowed us to photograph their items, with particular recognition to those who have done so anonymously. Further thanks to the Phonogalerie, Paris, for the pages from the 1903 Compagnie Française catalog, and Alison Rabinovici for her assistance with Stroviols. Our gratitude to Wim Prinssen for help with information about the Netherlands. For hospitality above and beyond the call of duty, we thank Stan and Suzann Stanford, George Glastris, Dennis and Rose Heindl, Steve and Tessie Oliphant, Jeff and Sandy Oliphant, Fabrice Catinot, Bernard, Marie-Louise and Jean-Paul Agnard, Yves Rouchaleau, John Levin, Garry and Susan James, and Gene and Helene Manno. Further thanks — Wyatt "Gesellschaft" Markus, Ed "Rambo" Fix, and Alan "Old Grease Collector" Mueller. And goodnight, W.C. Chin, wherever you are.

Chapter One

A Family of Phonographs

Discovery consists of seeing what everybody has seen and thinking what nobody has thought
— Albert Szent-Györgyi,
biochemist (1893-1986)

Boundaries

People commonly say, "It's a small world," yet recently the earth has seemed anything but compact. Not that it's getting any larger, of course (except in terms of population), instead it's growing more fragmented. Former empires such as the Soviet Union's have fractured into numerous minor sovereignties, and even modestly sized amalgams such as the former Yugoslavia have broken into much diminished pieces. Chauvinism, nationalism, civil war and xenophobia have been defining elements of the 1990s and the turn of the twenty-first century. Every day one is confronted with mistrust and misunderstanding from other cultures. So, perhaps it's the perfect moment to introduce to collectors who may have previously overlooked them, the similarities and harmonies that knit together the world history of talking machine companies.

In the talking machine's heyday, the globe was covered with far-reaching national empires. Great Britain and Russia immediately come to mind, but even less formidable powers such as Belgium had mini-empires. There were talking machine empires, too — Edison, Victor, and Columbia in the United States. The products these firms sold outside the United States were in fact quite similar to their domestic lines, yet American collectors have largely viewed exported artifacts as strange and mysterious. Collectors in other parts of the globe have understood United States instruments better than Americans have comprehended foreign machines, simply because the American presence in the history of recorded sound was so pervasive that a knowledge of American products would be impossible to avoid.

For example, most Canadian phonograph collectors could speak with understanding about machines manufactured in the United States. Canadians have always had collections populated by Edisons, Victors, and Columbias from across the border. Yet, until recently (following the publication of books such as our *Discovering Antique Phonographs* [2000], and Mark Caruana-Dingli's *The Berliner Gramophone*

[2005]), it would have been difficult to locate a United States collector who could tell you much about Canadian Berliner Gramophones, a brand that frequently encompassed products nearly identical to the Victor Talking Machine Company of Camden, New Jersey. These same American collectors could speak with authority about Victor, and show you the many Victors in their collections, yet it would have been rare for them to own any Canadian Berliners.

This is not an admonition for being small-minded, it is an observation of a phenomenon. In the United States, it has been "normal" to lack an understanding of foreign talking machines, even those of the neighbors to the north and south. Within the writing partnership of Fabrizio and Paul, there is one author highly interested in foreign instruments, and the other only marginally — so, claiming you "don't know anything about those foreign machines" is merely business as usual, at least in the United States. In spite of this, a delineation of the connections that existed between virtually identical talking machines sold in widely varying markets surely can benefit everyone, especially, with all due respect, the xenophobes.

Breaking Boundaries

The best illustration of global interconnectedness in the history of recorded sound is the Berliner/Victor/Gramophone story. It begins in Germany, from whence German-Jewish genius Emil (more frequently expressed "Emile") Berliner, a young man of 18 who had worked in a textile factory, immigrated to the United States in 1870. It was a time of turmoil in Europe; a time of war between France and Germany; a good time to get out. Moreover, Walter Kwiecinski reported in *The Talking Machine Review International*, No. 25, August 1973, that in the civic records of Berliner's hometown, Hannover, a handwritten entry read, "Secretly immigrated to America…," and the inventor was sentenced in absentia to a fine of 150 DM, or four weeks in prison, for evading military service.

After bouncing around a few jobs in his adopted country, Emile Berliner became interested in physics, and went on to design an improved telephone transmitter, which he sold to the Bell company in 1878. The money he received allowed Berliner the financial freedom to pursue further sound experiments — recorded sound.

Berliner drew inspiration from French inventor, poet, and visionary Charles Cros. Cros's 1877 sound-recording theory had proposed converting two-dimensional depictions of sound waves into a three-dimensional (and thereby "playable") format through photoengraving. In 1887, Berliner was convinced he could create a reproducible record from the spiral line traced on a carbon-coated glass disc by a stylus attached to a vibrating membrane.

Charles Cros had been a theorist, not an engineer, although he is viewed in France as the father of recorded sound. Yet, Berliner's experiments failed to vindicate Cros's photoengraving concept. Berliner could not obtain a satisfactory record by Cros's method. In Berliner's 1888 experiments, he succeeded in creating a replicable flat disc record by a three-dimensional process involving a wax-coated zinc disc. The zinc plate became a "master" record from which a mold or stamper was created that could press multiple, identical copies of a performance out of a "plastic" substance such as celluloid, hard rubber, or (finally) shellac.

From this discovery grew the "Gramophone" (writer of sound), as Berliner called his patented invention, a machine that played his mass-produced disc records (something the competing cylinder record format could not rival). It was all very brilliant, yet the Gramophone did not develop quickly. Around 1890, Berliner had his Gramophone manufactured as a toy in Germany. A ready market for the flimsy, hand-driven Gramophones was Great Britain, following the path of other German toy makers. The Gramophone's debut could not be described as a failure — yet how much success could be wrung from a simple plaything? Berliner set his sights on a bigger venture. In 1894, the Gramophone business began in the United States with the opening of a retail store in Baltimore, Maryland, offering a larger but still unsophisticated version of Berliner's hand-driven machine, and 7" diameter disc records.

The Berliner Gramophone Company was established in October 1895, but with little income to sustain it. In 1896, advertising man Frank Seaman contracted with Berliner to be his exclusive sales agent in the United States. Seaman's flair for people-oriented marketing sparked the sales of the $15.00 hand-driven Gramophones. A much-needed improvement, a lever-wound spring-driven version of the Gramophone, developed by mechanics Levi H. Montross and Eldridge R. Johnson, was introduced for Christmas 1896 at $25.00.

In mid-1897, the cabinet of the spring-driven Gramophone had been redesigned and given an improved soundbox devised by Eldridge Johnson and Alfred Clark. By August 1897, Johnson had revamped the motor, governor, and brake, and applied for a patent on what was to become known as the "Improved" Gramophone. This machine, selling for $25.00, finally would enable the Gramophone to become a viable competitor of cylinder talking machines. The "Improved" Gramophone would serve as a model for British academic painter Francis Barraud, who coupled it with a little fox terrier named Nipper, to eventually become one of the world's best-known trademarks: "His Master's Voice."

Just when everything appeared to be hunky-dory, dissention within the Berliner Gramophone organization broke it asunder. Frank Seaman was a businessman, a professional, a go-getter. He wanted Berliner to produce a lower-priced spring-driven model, something in the $18.00 range, capable of reaching a broader audience. He wanted Berliner to get Gramophones manufactured by the contractor who offered the lowest cost. Eldridge Johnson had the exclusive contract, and wasn't amenable to reducing the price he was charging. Certainly, Johnson was colluding with (and likely giving kickbacks to) certain directors of Berliner's company in order to maintain an unassailable position.

Seaman was also concerned about confusing terminology in the fast-expanding talking machine field. *Gramophone* sounded too much like *Graphophone*, a competing device. The consummate marketing man wanted a snappier and more easily distinguished moniker for Berliner's machines, to gain them continued attention. Emile Berliner, no doubt on the advice of men who were lining their pockets with the friendly assistance of Eldridge Johnson, acceded to his directors' rejection of Frank Seaman's innovative notions. Bad blood began to grow between the Berliner Gramophone Company and its merchandising magician.

It might have ended better if the marketing decisions had been left to an expert. Instead, the alienated Seaman, after several unfulfilled attempts to arrange the manufacture of less expensive, more efficiently designed Gramophones (as endorsed by his 1896 contract), finally decided to take matters into his own hands. In February 1898, Seaman formed the Universal Talking Machine Company. He hired a talented inventor, Louis Valiquet, who designed a high-quality disc talking machine, superior in several respects to the $25.00 "Improved" Gramophone, yet economical enough to be retailed for only $18.00. By the fall of 1899, Seaman had submitted the Valiquet machine for Berliner's approval, and had been rebuffed again.

And so, Seaman, through Universal, began to compete with the Berliner Gramophone. This expression of his frustration and disappointment branded Seaman as a scoundrel for decades, beginning with the publishing of the partisan distortions of Read and Welch's *From Tinfoil to Stereo* in 1959 until relatively recently, when the behavior of all the parties who contributed to fall of the Berliner Company finally began to be examined. It might well be said that a man of Seaman's verve could not endure the suffocating politics of the Berliner Gramophone Company.

Under the trade name "Zonophone," which he had earlier suggested to Berliner as a catchier handle than Gramophone, Seaman began marketing the Valiquet machine. In the spring of 1900, Seaman's Zonophone was licensed under American Graphophone (Columbia) patents, which gave the Columbia forces the "foot in the door" of the disc talking machine market that they had desperately wanted. Additionally, Seaman's firm, the National Gramophone Company, accepted a consent decree in court, thereby "admitting" as an agent of the Berliner Company, that American Graphophone patents were infringed by the Gramophone. American Graphophone quickly obtained an injunction against Berliner, effectively shutting down all Gramophone operations. Meanwhile, Columbia dealers began selling Frank Seaman's Zonophone.

Emile Berliner was rather suddenly unable to pursue his livelihood in the United States. His supplier, Eldridge Johnson, was at the time sitting on a newly-built factory full of machines that, by court order, could not be sold through traditional channels. Johnson grasped the only option open to him, though one of shaky legal standing — he became a retailer. With the help of another advertising whiz, Leon Douglass, Johnson attempted to sell Gramophones and records under his own name. Berliner could do little else but lend his blessing; he needed to regroup and reevaluate his position. Berliner focused on his existing facilities in Montréal, Canada, as a way to rebuild his Gramophone business.

Frank Seaman then brought suit against Johnson, alleging the Camden, New Jersey, business was merely a front for Berliner. Johnson, however, prevailed in court, and named his new corporation the Victor Talking Machine Company in recognition of his success. Emile Berliner resuscitated his Gramophone operations in Canada. The shoe, it might be said, was now on the other foot — Berliner came to depend on the largesse of Johnson. Emile Berliner was still the author of a formidable set of American patents, but these were put under the control of Victor, which used them to found a worldwide empire of immense proportions.

Before we examine the global impact of the Victor Talking Machine Company, let's take one last look at the man who more strenuously than anyone promoted the disc talking machine from its infancy. Frank Seaman's Zonophone was merely a stepping-stone in American Graphophone's march toward the introduction of its own Columbia Disc Graphophone. Columbia quietly withdrew its support of the Zonophone, and Seaman's sales agency went into receivership in September 1901. The following month, the first Columbia Disc Graphophones were introduced.

Seaman had proved he was a survivor before, and he tried it again. He reorganized his Zonophone business, introducing attractive new models throughout 1902 and 1903. Meanwhile American expatriate entrepreneur Frederick M. Prescott was operating a European branch of the business, offering some of the most fanciful instruments ever produced (see figure 3-60).

The International Zonophone Company was incorporated March 7, 1901 with capital of $40,000, and a registered office in Jersey City, New Jersey. Frederick Marion Prescott was one of the five stock subscribers. For some reason, the firm was reincorporated on May 6, 1901. Prescott was named managing director. He was to exploit the Zonophone in Great Britain, Europe and the Russian Empire. A factory and recording facility were set up in Berlin, Germany. At first, the International firm's products were virtually identical to the American branch's Types "A"-"D." Soon enough, the European catalog was brimming with inventive designs, and business was booming.

Back in the United States, however, Frank Seaman received something of a comeuppance. He suffered constant siege from the lawyers of vindictive Victor and fair-weather-friend Columbia. Finally, he packed it in, selling the Universal Talking Machine Manufacturing Company to Victor in September 1903. This transaction was directly linked to the 1903 acquisition of the International Zonophone Company by Deutsche Grammophon, the German subsidiary of the Gramophone and Typewriter, Limited (both Victor affiliates). Zonophone, already a significant force in the industry, would continue to be an important and enduring brand, though Frank Seaman's contribution to the history of recorded sound had ended.

Thus was the fate of the Gramophone decided in the Americas. Berliner's invention had been the first and only successful disc talking machine the United States, but that exclusivity was ended, replaced by Berliner's sanctioned successor Victor versus Zonophone (while it remained independent), Columbia, and whatever others could fight for survival in the courts and the marketplace. Emile Berliner would henceforth control the Canadian business.

Victor, beside the United States, would exploit the South American market, in which Argentina and Uruguay appear to have been the most significant outlets.

Beyond Boundaries

By the time of Berliner's troubles with Frank Seaman, the Gramophone had been well established abroad. In mid-1897, William Barry Owen, a director of Seaman's Gramophone distribution company, had secured the patent rights to the Gramophone in Great Britain and Europe. Owen arrived in London seeking investors to help him get established there. For almost a year, his efforts at persuasion failed to attract any British capital.

Finally, in May 1898, Owen assembled enough investors to form the Gramophone Company, Limited, the purpose of which was to tap the market in the many countries where Berliner's invention was virtually unknown. Offices and a recording studio were set up in the basement of 31 Maiden Lane. Owen was named managing director, while Emile Berliner's brother, Joseph, was sent to Hannover, Germany, to establish a record pressing plant (which marked the beginning of Deutsche Grammophon).

The Gramophone Company, Limited was the seed from which a European network of "Gramophone Companies" would grow. The currents that flowed within this network may be recognized as follows. Great Britain's Gramophone Company (between the end of 1900 and the end of 1907, the Gramophone and Typewriter, Limited, during which the firm engaged in a rather unsuccessful exploitation of the Lambert typewriter) was the parent firm, the hub from which merchandise and parts arriving from the Victor factory in Camden, New Jersey, moved into Europe. The Continent was effectively divided into Gramophone territories. For instance, the Spanish branch was owned by the French branch, and the German branch handled points east (the Great War would disrupt Deutsche Grammophon's direct ownership by London). The Italian agency operated as the Gramophone Company (Italy), Limited.

London's affiliate in the Netherlands had an interesting history. Frederik Willem Jambroes von Bemmel Wortman was operating the American Import Company in Holland when the Gramophone European territories were laid out in 1899. Wortman already had imported Gramophones and records from the Berliner Gramophone Company. In consideration of this, Wortman's existing enterprise was converted to an agency for the Gramophone Company, Limited, covering the Netherlands, Belgium and "the Colonies." Wortman's firm retained some of its autonomy, a quality which allowed it to arrange for recordings in local dialects. The renowned recording expert Fred Gaisburg contributed to this project, but, curiously, so did another American, Cleveland Walcutt, who had worked for Edison in the 1890s and, by the early years of the twentieth century, had found his way into the employ of Compagnie Française du Gramophone. Although the discs were local, the Gramophones followed a pre-established path beginning in Camden.

In fact, during the first decade of the twentieth century, a large portion of the products sold by these far-flung firms originated in Camden, New Jersey. As the decade progressed, the cabinetry of the British and European Gramophones began to vary markedly from what was being offered in the United States. European cabinets were frequently far more artful and imaginative than those of American Victors. As a side note, it is interesting to observe that Victor instruments sold in South America were identical to those in the United States catalog. In Great Britain and Europe, however, the dictates of cultural taste required that many of the cabinets be produced locally. Even when this occurred, mechanical parts continued to be imported from Camden.

Meanwhile, in Montréal, Québec, Canada, Emile Berliner was pursuing a different path. Berliner had been granted Canadian patents for his Gramophone in 1897. In 1899, he opened a retail store in Montréal. Canadian law required that to maintain his patents, Berliner needed to begin production in Canada, so he started assembling Gramophones there from mechanical parts imported from the United States, employing locally made cabinets. In 1900, Berliner began pressing records, on a modest scale, in Canada. When the Columbia injunction stymied his United States operations, Berliner's move to Canada (although his business relocated there, he continued to live in the United States) was a logical extension. His Canadian branch, like the Gramophone Company, Limited of London, was already established, and Montréal was convenient by rail to Eldridge Johnson's Camden factory.

Although Johnson was "victorious" in 1901 against Columbia's efforts to enjoin him out of existence (which might have opened the door to a renewed Berliner presence in the United States), Emile Berliner decided to put all his personal efforts and those of his family into building up the Canadian trade. He owned stock in Johnson's newly-formed (October 3, 1901) Victor Talking Machine Company, and he continued to hold a close relationship with Victor that allowed him to share in on-going innovations such as the "Taper Arm."

Whereas it took years for the British and French Gramophone affiliates to develop their own distinctive products,

Berliner started off on a very personal path in Canada. Instead of relying heavily on imported Victor parts and cabinets, Berliner moved rapidly to manufacture virtually everything he needed in Montréal. The Canadian Model "E" (second version) of 1905 (see *Discovering Antique* *Phonographs*, Fabrizio & Paul, figure 3-65) outwardly resembled a Victor "Royal." However, disassembling the Canadian machine reveals that it was manufactured by means similar but not identical to Victor's Camden factory.

1-1. No, it's not the step pyramid of King Djoser — it's an early version of the Canadian Berliner Model "E," circa 1903. This instrument definitely demonstrates how distinctive were the designs Emile Berliner exploited in Canada. The triangular cabinet configuration did not catch on in the United States, but was represented in Canada and Europe. *Courtesy of Richard and Jill Pope (Value code: G)*

1-2. Hmmmm… this looks familiar. Over the years, the German catalog of Deutsche Grammophon, included at least two "pyramidal" models. The one shown here incorporated pressed wooden sides (as also seen in figure 3-93). Another type is pictured in *The Talking Machine, an Illustrated Compendium*, Fabrizio & Paul, figure 3-133. The Gramophone affiliate serving the German territory offered imaginative designs that deviated from the Victor paradigm earlier than G&T or Compagnie Française. *Courtesy of Jalal and Charlotte Aro (Value code: F)*

So it was that Emile Berliner's Canadian enterprise followed the development of the British and French branches in reverse. Until 1910, Berliner produced a specialized line of Gramophones in Canada, generally similar to Victors, but sometimes highly original, such as the early style "E" seen above (or the "J," see *Discovering Antique Phonographs*, Fabrizio & Paul, figure 3-67). After 1910, Berliner ceased the manufacture of Gramophones, and became a hugely successful Victor distributor. Imported Victor instruments were affixed with a little plate identifying the Canadian firm.

A Monarch by Any Other Name

It so happened that Leon Douglass, Eldridge Johnson's marketing expert in the early days, chose a couple of "old world" terms to describe Victor's post-Berliner talking machines. "Royal" was one of the names he used, but a far more significant epithet was "Monarch." Although "Monarch" initially referred to only the biggest and most expensive of Victor's machines and records, it clicked with the public and was soon more broadly applied, as in "Monarch Junior."

The word "Monarch" resonated in Great Britain and Europe, too (in German, for instance, "Monarch" is spelled the same as in English, and the term was vigorously exploited by Deutsche Grammophon). As a merchandising message it worked well wherever it was used, even in countries such as the United States and France, which had fought revolutions against the rule of kings. Pathé may have flattered egalitarians with its "Democratic" cylinder phonograph, but the French public, perhaps with a touch of nostalgia, readily accepted Victor's "Monarch" ("Monarque," in French).

A study of Victor's French affiliate, Compagnie Française du Gramophone, excellently illustrates the influence of Camden in what many Americans would consider an exotic locale — and shows the strong "family resemblance" of all Victor/Gramophone products. Let's examine the October 1903 Compangnie Française catalog in detail.

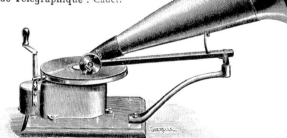

— 1 —

GRAMOPHONE N° 3.

Bien qu'il soit notre plus petit modèle et d'un prix minime, ce **Gramophone** donne d'excellents résultats.

Il est muni d'un grand ressort et peut donc jouer indistinctement : disques " **Monarques** ", disques " **Concerts** " et petits disques.

Son mécanisme est protégé par une boîte en bois verni, ou par une boîte en métal nickelé. Le client devra indiquer quelle boîte il désire; dans le cas contraire, nous enverrons à notre gré.

L'appareil est livré complet prêt à fonctionner, muni d'un pavillon nickelé et d'une boîte contenant 200 aiguilles.

APPAREIL BOITE BOIS

Dimensions : 55×25×44.
Poids : { brut 10 kil.
{ net 4 kil. 250
Code Télégraphique : Abies.

APPAREIL BOITE NICKELÉE

Dimensions : 55×25×42.
Poids : { brut 10 kil.
{ net 4 kil. 800
Code Télégraphique : Cadet.

PRIX : **62 fr. 50**

1-3.

Even in late 1903, the "Improved" Gramophone designed by Eldridge Johnson for Emile Berliner in 1897 was still being offered. Remember, this model (frequently called the "Trade Mark" today, because of its association with the "His Master's Voice" logo) had been the very first spring motor disc talking machine produced in significant numbers. It was primitive by comparison with the more "modern" instruments of 1903. Following the Berliner/Columbia debacle, Eldridge Johnson had sold the "Improved" or "Trade Mark" in the United States during 1900, and dressed it up a little in 1901 (the Model "B"). Yet, the plain old "Trade Mark" continued to be shipped abroad for years thereafter (shown at left of figure 1-3).

In 1904, Compagnie Française sale-priced the "No. 3" "Trade Mark" instrument, knocking off ten francs, depending on the type of soundbox included, as the iconic

Berliner apparatus was finally phased out in Europe. There is evidence of cost-reduction as the apparatus was slowly retired — at least one collector reports a motor cast from inexpensive "pot-metal" rather than stable iron or brass.

The machine on the right in figure 1-3, which Victor called the Model "A," was a transitional product in the United States, bridging the period between Eldridge Johnson's eponymous line of 1900, and the formation of the Victor Company (October 3, 1901). In France, the Model "A" had already undergone a formidable price reduction by the time it appeared in this 1903 catalog — in 1901 Compagnie Française had listed it for a considerable 100 Francs! Both machines had been retired from the American market by 1902, but endured, as evidenced in figure 1-3.

1-4. Deutsche Grammophon, the German affiliate, sold this "Trade Mark" machine, somewhat different in appearance than those of the British and French Gramophone branches. The lower cabinet molding is not quite the same, and the decal is a considerable departure. *Courtesy of Ira and Dottie Dueltgen (Value code: D)*

1-5. A decorative emblem, rather than a simple rectangle, the decal is difficult to decipher, but listed one of Emile Berliner's German patents: No. 45048. Note, stamped on the surface of the bottom molding, "NOT LICENSED FOR SALE IN FRANCE OR GREAT BRITAIN." The fact that this admonition is written in English probably indicated the pecking order of the European Gramophone network, with the Gramophone Company, Limited (or G&T) at the top. *Courtesy of Ira and Dottie Dueltgen*

It's a bit ironic that the one instrument in the Compagnie Française catalog that exploited the "Monarch" terminology (the "Monarque Junior" shown here) was akin to a Victor "P" ("Premium"), whereas certain machines sold in France under simple number designations were known as "Monarchs" in the United States. Considerably different in appearance than the model seen here was the instrument that went by the name "Monarch Junior" in the American catalog (and was never sold outside the United States). Yet, the nomenclature finally took hold in France. By 1908, practically every instrument in the French roster was called by some variation of the "Monarch" name.

1-6.

1-7.

American collectors will recognize the characteristically baroque styling of the Victor "MS" ("Monarch Special") in the Gramophone "No. 7," shown in figure 1-7. What isn't evident from the drawing is that this particular version employed a one-spring motor. The "Monarch Special" was a deluxe instrument in the United States, having a robust, three-spring motor. In Great Britain and Europe, it was offered with either single or triple springs. The lesser version, shown on the catalog page at left, did not have a hinged top-board, so access to the motor was made by removing the bottom of the cabinet. It employed the same motor Johnson had sold during 1901 in his first "Monarch" machine, generally called the "Johnson Monarch" (an instrument that had been sold in the United States, Great Britain and Europe).

1-8. This is the "Johnson Monarch" as it was sold in Great Britain and France (Gramophone "De Luxe No. 9"). It is, by all appearances, the same instrument sold in the United States, but instead of the conventional Victor metal ID plate, it carried a rectangular decal as seen here. *Courtesy of Bert Gowans (Value code: F)*

1-9. The motor of the 1901 Johnson "Monarch" and Gramophone "De Luxe No. 9" was then used to drive the 1903 Gramophone No. "7." Johnson had long since recovered his set-up costs for this one-spring motor, so exporting them in significant quantities must have been a lucrative business. The 1904 Compagnie Française catalog listed the No. "7," configured with the improved "Taper Arm," for 200 FF, the same price asked for the front-mounted version (see next illustration). *(Value code: F)*

1-10. A "Taper Arm" Gramophone No. "7" equipped with a horn of extraordinary size. The red interior was a popular feature of European horns of the period, most commonly associated with Zonophone or Odeon. *(Value code: E)*

1-11. Until late in 1904, Victor did not offer any talking machines in mahogany cabinets in the United States. In Great Britain and Europe, however, mahogany was considered a stylish alternative to utilitarian-looking oak, and the "Trade Mark" was available in a mahogany cabinet (called the "Mozart" in France) in 1901. In 1903, the presently discussed Gramophone No. "7" could be special ordered in mahogany, an example of which is seen here. *Courtesy of Brice Paris (Value code: VR)*

1-12. Having just said that Victor did not make mahogany instruments before 1904, we present this highly unusual Victor "MS" custom mahogany cabinet from 1902. Evidently for the right price, Victor *could* be induced to abandon its practical oak. *Courtesy of Lynn Bilton (Value code: F)*

1-13. Speaking of interesting variations… In 1900-1901, while Eldridge Johnson was marketing his Victor Type "C" in the United States, the same mechanism was being shipped abroad, to be housed in the cabinets of British and European Gramophones. To the left is shown the earliest of these instruments, sold by the Gramophone Company, Limited, in 1900, just prior to firm's name change to the Gramophone and Typewriter, Limited. Note the decal reflected the style and shape of the Berliner-era. Later G&T and Compagnie Française decals were long and narrow. *Courtesy of Richard and Jill Pope (Value code: F)*

1-14. See caption for figure 1-15. *Courtesy of Bert Gowans (Value code: F)*

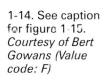

16

1-15. The Victor "C" mechanism was also incorporated into the "old numbered" Styles "6" and "7." "Old numbered" indicates that the same model numbers were later reused for different instruments as product lines developed and newer designs appeared. The 1901 No. "6" (figure 1-14) and No. "7" (figure 1-15) both had a metal motor plate surmounting the cabinet, and a round horn support arm, which were characteristics of Victor's Type "C." This particular No. "7" had a Compagnie Française decal laid over a G&T decal, leaving only the left side, with an image of the "Recording Angel," exposed. This indicates as well as anything the way Victor products flowed to Europe. The Victor "C" mechanism left Camden, arrived in Great Britain, was enclosed in a stylish cabinet, and then was shipped to France. *Courtesy of Jalal and Charlotte Aro (Value code: E)*

– 7 –

GRAMOPHONE N° 9

Dimensions : 74×33×54.

Poids : { brut 21 kil.
{ net 8 kil. 550.

Code Télégraphique : Avona.

Tout en réunissant les avantages du **Gramophone** N° 7, notre N° 9 lui est cependant supérieur sur plusieurs points.

Son grand avantage consiste dans le mécanisme ainsi que dans le ressort dont la puissance a été doublée, ce qui lui permet de jouer en moyenne, lorsqu'il est remonté à fond, 4 petits disques ou la quantité correspondante de disques concerts ou monarques.

En pressant sur le bouton qu'on aperçoit à droite de la manivelle, on peut soulever le couvercle sur lequel est adapté le mécanisme, et se rendre compte de celui-ci.

Il est livré complet, prêt à fonctionner, muni d'un pavillon cuivre nickelé, grand modèle, et d'une boîte contenant 200 aiguilles.

PRIX : **250 francs**.

1-16.

American collectors might be puzzled by the catalog page in figure 1-16. Shown is a machine identical to the Victor "Monarch," sold 1902-1905 in the United States. The puzzling aspect is that in France this instrument was given a higher place (and a higher price) in the roster than the preceding "No. 7" Gramophone, which was the veritable spit and image of an American "MS." In the United States, the "MS" was a step above the "Monarch." The key, however, lies in the motor. The "Monarch" (or new "No. 9") shown here had two springs, hence it occupied a loftier place in the hierarchy, despite a less ornate cabinet.

Gramophone "No. 13" was the same Victor "Monarch Special" that American collectors know. Only the horn varied slightly — nickel-plated brass, in keeping with contemporary European taste, rather than plain brass or steel body/brass bell. The following year, a "Taper Arm" version of the "No. 13" was offered in the French catalog, at no increase in price. As Victor's talking machines evolved, the design configuration of the horn and arm(s) underwent significant changes. The instruments shown above encompassed the early "straight" horn arrangement, which used two arms for support. In October 1902, Victor announced its improved "Rigid Arm" apparatus, seen next.

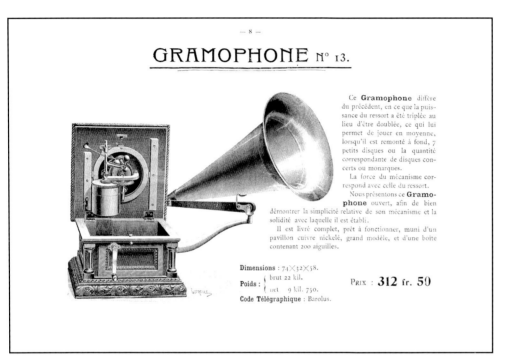

1-17.

1-18.

The "Rigid Arm" relieved the record of the inertia of the horn (thereby achieving less record wear), which was positioned over the top of the cabinet in a space-economic position. In the United States, the "Rigid Arm" began to be supplanted by the more acoustically correct "Taper Arm" after only six months on the market. For export, however, the "Rigid Arm" was available even in 1904, offered concurrently with the "Taper Arm" (the 1904 Compagnie Française catalog listed the "Rigid Arm" attachment separately at 50 FF, to adapt a front-mounted Gramophone).

The advertising copy in figure 1-18 rivals any pseudo-scientific piffle dished out by a talking machine firm, "The soundbox is able *at the same time* to move horizontally and vertically, and this with an absolute precision. Sounds travel through the steel tube [in actuality it was brass], where they are refined, to exit completely pure through the horn, moveable on its axis, which permits it to be oriented in the direction most favorable for the sound or the placement of the listeners, without which the position of the Gramophone would need to be changed."

1-19. The French and British catalogs included this interesting model, which was never sold in the United States. Known as the Gramophone No. "4a," it was distinguished by a motor somewhat similar to the "Improved Gramophone" or "Trade Mark," set diagonally within an oak cabinet. The turntable was 10" in diameter, larger than the "Trade Mark," but the winding crank of the "4a" retained the "old fashioned" vertical position. It was equipped with "Rigid Arm." Note the speed control emerging from the front of the cabinet at a steep angle, caused by the diagonal placement of the motor. Identifying model numbers tended to change as new catalogs were issued, but the mechanism associated with the No. "4a" was sold both in a plain cabinet, and an embellished one with beading and beveled sides (seen here). *Courtesy of Richard and Jill Pope (Value code: D)*

1-20. In the United States, Victor's "Rigid Arm" attachment was a brilliant flash of innovation, rapidly superseded by the "Taper Arm," a design that was to endure for decades. In Great Britain and Europe, however, the Rigid Arm continued in use at least a year after it was discontinued in the United States. It would be easy to apply the "robber baron" hypothesis here, and accuse Victor's Eldridge Johnson of "dumping" obsolete parts into the foreign market. Yet, the situation is more complex than can be satisfied by an explanation as simple as that. America was (and to a certain extent remains) a land of innovate, binge, and discard. Great Britain and Europe, on the other hand, tended to create traditions, to feel more comfortable with familiar and acceptably functional objects. The rarest variation of the Gramophone "4a" is seen here — in mahogany finish, unusual even in Europe where mahogany was more popular than in the United States. *Courtesy of Jalal and Charlotte Aro (Value code: VR)*

— 10 —

GRAMOPHONE Nº 15

Dimensions : 75×35×58.

Poids : { brut 24 kil.
{ net 10 kil. 800.

Code Télégraphique : Gigas.

Cette machine est la plus puissante que nous ayons. Elle est munie de trois ressorts qui lui permettent une audition considérable **sans être remonté.** — Sa boîte est très large, joliment sculptée, et le plateau qui reçoit le Disque a un diamètre de 304 ᵐ/ₘ.

Cet appareil est donc, par sa force et son importance, spécialement construit pour jouer nos Disques " **Monarques** ", dont la durée d'audition peut aller à quatre minutes environ et leur force en proportion de leur surface. Naturellement il joue de mêmes des disques concerts et petits.

PRIX : **375** francs

1-21.

Gramophone "No. 15" (figure 1-21) could easily be mistaken for an "MS," yet the written specifications reveal the cabinet to be slightly larger in size. In fact, this was a Victor/Gramophone product never offered in the United States. It was a "straight horn" (front-mounted) version of the instrument sold to Americans as the Victor "D" (equipped with "Taper Arm"). The "D" was not available as a front-mount in the American market. The cabinet of the Victor "D" imitated the "MS," but was somewhat larger to accommodate a wider (12" diameter as opposed to 10") turntable. The three-spring motor was the same as an "MS," except it was nickel-plated.

In the United States, the so-called "fancy" Model "D" (the cabinet was subsequently simplified) is presently considered a rare machine. It was top-of-the-line in late-1903 and 1904, but observation of extant examples suggests that the modest increase in the size of the turntable was not sufficient to persuade clients to choose it over the similar "MS." In Europe, however, the "D" was a greater success; it was sold in the straight horn version shown here, and later in the back-mounted "Taper Arm" configuration as seen in the United States. In this form it remained available from Victor foreign affiliates, the Gramophone and Typewriter, Limited, for example, until about 1908, by which time it had been adapted to receive the improved Victor worm gear motor. G & T called it the "Senior Monarch."

The 1903 Compagnie Française catalog contained this extraordinary bit of engineering, called "Le Meuble-Support" (record cabinet support), 250 FF (not including the body of the talking machine). "… The moldings executed on the record stand harmonize marvelously with our apparatus [illustrated is a "No. 15"]. Equally, the price of the "Meuble-Support" is found to include the price of a large brass horn, the combined dimensions of which allow discs to be appreciated outdoors at a distance of more than 500 meters." Deutsche Grammophon advertised the same arrangement of cabinet with huge horn attached in May 1902, as the "'Piedestal' mit Demonstrationstrichter" (stand with exhibition horn). It's interesting to note that Carl Below, a Leipzig manufacturer with well-established ties to Deutsche Grammophon, had advertised a very similar cabinet/horn configuration three months earlier (in the *Phonographische Zeitschrift* trade magazine).

— 15 —

MEUBLE ⸺ SUPPORT

De construction très élégante, ce meuble a été construit de façon à ce que le pavillon se trouve dans les meilleures conditions exigées par l'acoustique. Ainsi qu'on en peut juger par la figure ci-contre, les moulures exécutées sur le meuble s'harmonisent à merveille avec notre appareil.

Dans le prix du Meuble-Support, se trouve également compris le prix du grand pavillon en cuivre, dont les dimensions combinées permettent d'entendre les disques en plein air à plus de cinq cent mètres de distance.

Toutefois, le même pavillon sert également pour auditionner dans un salon, sans qu'on ait à craindre une trop grande résonnance.

C'est là un immense avantage que nous tenons à signaler spécialement.

PRIX : **250 francs.**

Code Télégraphique : Agrone.

Le pavillon avec lyre : **75 francs.**
Code Télégraphique : Carcic.

Le pavillon avec lyre et trépied : **90 francs.**
Code Télégraphique : Corusi.

1-22.

A World of Antique Phonographs

It must be said, then, that the instruments sold by Eldridge Johnson/Victor in the United States, and those sold by Compagnie Française were essentially the same during the 1901-1904 period. Zonophone, after it lost its independence in 1903 and became an adjunct of the Victor/Gramophone affiliates, offered machines in both North America and Europe that ever increasingly lost their Frank Seaman-era look, and evolved into cut-rate versions of the mother brand's line.

The other American talking machine firms doing business in Europe at this time, Columbia and Edison, made fewer accommodations than Victor to local taste. The instruments they sold outside the United States were identical to their domestic issues except for horns. Edison Phonographs, for instance, were sold in Great Britain and France with spun aluminum horns, local favorites. Columbias could be had in Great Britain with small, unsupported flower horns. A 1907 *Talking Machine News* advertisement from Columbia's London office offered the Type "BV" Graphophone under the name "Trump," equipped with the aforementioned miniature flower horn, an accessory seldom seen in the United States.

Of course, there were also unauthorized British and European copies of American instruments. Some of these were one-of-a-kind "science projects" scrupulously created by skilled machinists, others were "catch me if you-can" knock-offs. Pathé in France blatantly adopted the designs of Edison ("Le Gaulois") and Columbia ("Le Français," Le Coq," "Le Stentor" and others). Edison Bell in Great Britain produced somewhat simplified copies of Edison Phonographs as part of an agreement with Edison that the American company came to regret. The Germans were especially adept at imitating the work of Victor, Edison and Columbia. Even this back-handed "homage" must be seen as another thread of connectivity between firms on either side of the Atlantic. Collectors cannot, wherever they live, escape evidence of the world network of technology that created the antique phonographs they admire.

1-23. A rare example of the Compagnie Française "Meuble-Support." The record cabinet was ebonized, a posh finish at the time. *(Value code: VR)*

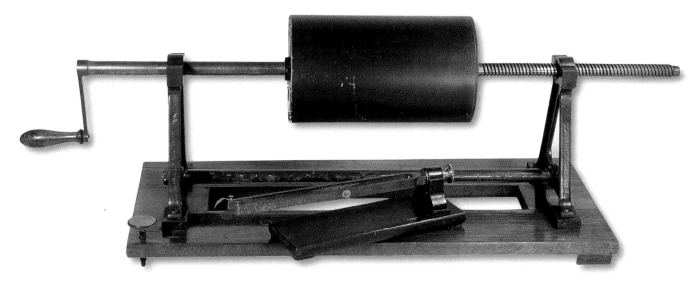

2-1. A Phonautograph apparatus built by instrument maker J. Lancelot of Paris. The original 1857 design of Leon Scott de Martinville employed the lamp-blacked surface of a disc upon which tracings of sound waves were registered. By 1859, Scott had abandoned the disc design in favor of a cylinder, as shown here. By turning the crank, the coarse threads to the right caused the entire cylinder and shaft to traverse horizontally. Various arrangements of styli could trace patterns of sound waves on the turning cylinder. The smaller device on a black wooden base is an adjustable tuning fork for use with this Phonautograph. See next illustration. *(Value code: VR)*

2-2. A close up of the tuning fork attachment in place on the J. Lancelot Phonautograph. The small sheet brass stylus on one arm is probably not original. In actual use, a hog's bristle would have been attached to each arm to register the vibrations of the tuning fork onto the delicate lampblack coating of the cylinder.

2-3. The earliest talking machines (1877-1885) employed sheets of tin foil as the recording medium. These records were not permanent, becoming effaced after only 2 or 3 recitals. Nor were they very loud or distinct; it was noted at the time that being present when the recording was made greatly aided in its recognition during playback. Yet, for a world newly awakened to the possibility of recorded sound, the phonograph seemed miraculous. This unmarked example measures 9 1/2" x 16". Certain features such as subtle cross-hatching on the brass speaker arm suggest that this is a mid-1878 Bergmann reworking of a slightly earlier Bergmann or Patrick & Carter exhibition instrument. *(Value code: VR)*

2-4. This remarkable image from a stereo card captured a Phonograph exhibition circa 1878. The machine is similar to the example shown in the previous illustration except for the unusually large flywheel. This feature was an improvement to better regulate the hand-driven operation. *Courtesy of David Giovannoni (Value code: VR)*

2-5. A vintage photograph reveals a substantial "Tin Foil" Phonograph manufactured by Alex Pool with advanced features. Note the reproducer/recorder mounted to a long swing-arm, the massive crank/flywheel to the right, and feedscrew disengagement ("throw-out lever") at far left. Pool made ten of these instruments for Edison, who had them modified by Charles Batchelor. The throw-out lever would soon be repositioned beneath the shaft, and a new speaker arm that tilted away from the mandrel would replace the cumbersome component shown here. *(Value code: VR)*

2-6. The obverse of the photograph supplies valuable details, not the least of which is the exact date of the image. The "Charles P. Edison" who signed the photo was Thomas Edison's 18 year old nephew (known to all as "Charley") who became an assistant at the Menlo Park lab, and the inventor's unofficial photographer. Brilliant but erratic, Charley made significant contributions to Edison's work, and there was widespread belief that he might well follow the footsteps of his famous uncle. Charley, however, died in Paris the following year of peritonitis.

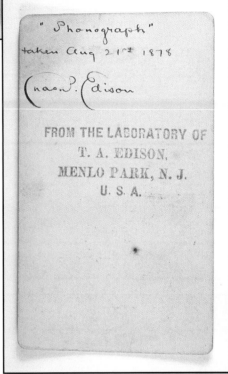

2-7. A rear three-quarter view of the same remarkable Pool "Tin Foil" Phonograph shows details of the feedscrew disengagement at right, and what appears to be an improved tin foil securing device over the cylinder. By the fall of 1878, Edison's attention would be diverted to developing a practical electric light and a power generating/distribution system, and his days of "Tin Foil" Phonograph experimentation would cease. *(Value code: VR)*

2-7a. A seemingly unremarkable view of a section of Main Street, Springfield, Massachusetts. Close examination, however, reveals decorations in place for a massive Grand Army of the Republic reunion, which is documented by the *Springfield Republican* to have occurred on June 5 and 6, 1878. On page 6 of the June 6, 1878 edition, a small item stated, "Edison's phonograph is again on exhibition in Hampden Hall, having returned from a short visit to Northhampton. It remains here the rest of the week." See next image. *Courtesy George G. Glastris*

2-7b. A close-up of the previous image reveals a banner over the entrance to Hampden Hall, which reads: **"EDISON'S PHO-NOGRAPH - Talks, Laughs, Sings &c. - ON EXHIBITION - 10, 12, 2, 4, 8, 10 - AD-MISSION 25c."** *Courtesy George G. Glastris (Value code: VR)*

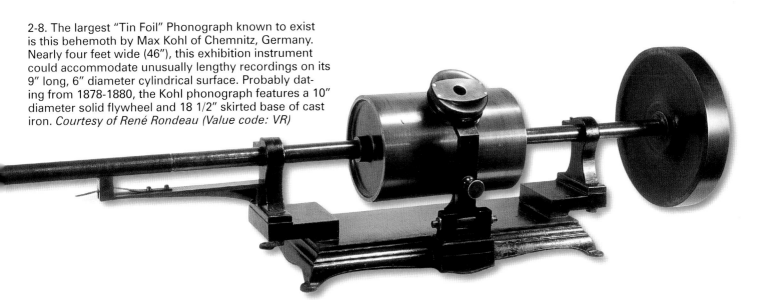

2-8. The largest "Tin Foil" Phonograph known to exist is this behemoth by Max Kohl of Chemnitz, Germany. Nearly four feet wide (46"), this exhibition instrument could accommodate unusually lengthy recordings on its 9" long, 6" diameter cylindrical surface. Probably dating from 1878-1880, the Kohl phonograph features a 10" diameter solid flywheel and 18 1/2" skirted base of cast iron. *Courtesy of René Rondeau (Value code: VR)*

2-9. Beneath a gleaming brass mandrel, Max Kohl's marking on the base of his exhibition "Tin Foil" Phonograph. *Courtesy of René Rondeau*

2-10. The French firm of Ducretet & Compagnie manufactured "Tin Foil" Phonographs for educational use into the 1890s. The firm's machines closely resembled those of Urbain Marie Fondain, whose business passed to Ducretet in 1881. Most of Ducretet's instruments were not equipped with a flywheel, which makes this particular example unusual. *Courtesy of René Rondeau (Value code: A)*

2-11. The first examples of the venerable Edison Class "M" Phonographs to appear in 1888 were configured like this one. Most noticeable is the "spectacle device" that allowed rapid interchange between the reproducer and recorder. Note also the small secondary belt off the main shaft which drives a carriage return screw. Both of these features had a short life and had disappeared by November 1889. The wooden swarf drawer (to catch wax shavings) below the brass mandrel is missing from this example. *(Value code: VR)*

2-12. This cabinet photograph (4 1/4" x 6 1/2") of Thomas Edison was taken by Daireaux in Paris, possibly during Edison's visit during 1889. By that time Edison was well-known as "The Wizard" and the photographer might have envisioned a large demand for the inventor's likeness. It is an unusual pose for Edison, who usually stared directly into the lens of the camera. *(Value code: VR)*

2-13. Among the 1891-1895 motor-powered Graphophones (Types C, E, F, I, K, R, S, U) that employed (originally treadle-driven) 1888-1890 Bell-Tainter playing mechanisms (Types A, B, C), the Type "C" was a bridge between the two eras. Equipped with both a treadle mechanism *and* an electric motor, the Type "C" ("Combination") was meant to remain functional in spite of any caprice of early storage batteries. Shown here is an example of a Type "C" which conforms to no known type marketed by American Graphophone. This machine displays very interesting period modifications and contradictory clues to its real origin. *Courtesy of the Gfell family collection (Value code: VR)*

2-14. This view shows the Type "C" pulley plate with the serial: No. 10226. A study of known serial numbers would suggest that this playing mechanism was originally marketed as a Type "C" treadle machine — not the slightly later Type "C" "Combination" Graphophone. One problem with this hypothesis is the presence of a wide, channeled pulley rather than a narrower model with a V-shaped groove as employed on treadle machines. It appears that this mechanism was refitted with a later, wider pulley, but retained its original identification plate. *Courtesy of the Gfell family collection*

2-15. A view of the motor suggests that whoever modified this Graphophone was inspired by the excellent Edison bi-polar electric motor that powered the Edison Class "M" Phonograph. A four-ball governor (like that found in electric motor Graphophones) was employed in a vertical position, much like the two-ball governor of the Class "M." The execution of this specimen was well designed, professional, and efficient. In light of this and the presence of a later wide pulley with an early serial numbered plate, it seems likely that this machine was recycled from a parts inventory — suggesting a large retailer such as the Chicago Talking Machine Company, or perhaps the Graphophone factory itself. *Courtesy of the Gfell family collection*

2-16. The 1894 Type "G" Graphophone ("Baby Grand") represented American Graphophone/Columbia's transition from wringing survival out of the 1891 collapse of their business prospects by cobbling together outmoded "Bell-Tainter" cylinder talking machine mechanisms, to manufacturing newly-inaugurated apparatus. The "G" very much resembled the ill-fated` "Bell-Tainter" devices, but it had one revolutionary feature: a convenient and reliable spring motor designed by Thomas Macdonald. The significance of this instrument in the transformation of American Graphophone from abject failure to formidable rival of Edison can be attested by the metal tag affixed to it. Columbia kept an archive documenting the Graphophone's history at the American Graphophone factory in Bridgeport, Connecticut. Artifacts from this collection were exhibited in the Liberal Arts Building of the 1904 St. Louis World's Fair. In the official program, this machine was listed as "4. Original model Macdonald Spring Motor, the first spring motor used on a talking machine." This claim of primacy was a distortion (see *The Talking Machine, an Illustrated Compendium*, revised, Fabrizio & Paul, figures A-7 – A-10), but Columbia was in the business of selling Graphophones, not teaching history. *Courtesy of Sam Sheena (Value code, ordinary Type G Graphophone: B)*

2-17. From 1891 until 1894, the Edison Phonograph Works manufactured cylinder blanks referred to as "channel rim" by modern collectors. Testimony by Walter Miller, the manager of Edison's record plant, strongly suggests that this design was meant solely for duplicated recordings (rather than "originals" or "masters," directly recorded). Examples do exist with ring-shaped paper labels in the channels carrying printed titles and catalogue numbers, but many more are found with no labels, as seen here. (Some labels may have become detached from their channel-rim cylinders, yet home recordings also exist in this format, and blanks sold for such use would not have been equipped with labels.) When the firm of Walcutt & Miller assumed operation of the Edison record plant in 1895, the channel rim remained, although no W&M ring labels are known. During the channel-rim cylinder's brief, erratic life, variations in color and length have been noted, but the unusually short example shown to the left of a typical size cylinder is an oddity. Most users felt that the standard 4" length cylinder was constraining enough. Why these blanks were offered in truncated size is a mystery. The recording on the short cylinder is an excerpt of a poem by American poet Fitz-Greene Halleck (1790-1867) entitled *Marco Bozzaris*. It concerns a Greek patriot (actual name Marko Botsaris) who struggled against the Ottoman Turks in the early 1820s. *At midnight, in his guarded tent,/ The Turk was dreaming of the hour/ When Greece, her knee in suppliance bent,/ Should tremble at his power… Courtesy of the Scott and Denise Corbett collection (Value code: VR)*

2-18. In the late 1880s and early 1890s a few designers offered record boxes designed to protect the fragile wax cylinders during transport. These improved designs usually incorporated an additional stabilizer in the lid to prevent the wax from contacting the interior sides of the box. One of these early designs is shown here, along with the recording it has protected for over a century: "The Whistling Coon," by George Johnson, the first successful African-American recording artist. *Courtesy of the Scott and Denise Corbett collection (Value code: VR)*

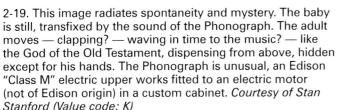

2-19. This image radiates spontaneity and mystery. The baby is still, transfixed by the sound of the Phonograph. The adult moves — clapping? — waving in time to the music? — like the God of the Old Testament, dispensing from above, hidden except for his hands. The Phonograph is unusual, an Edison "Class M" electric upper works fitted to an electric motor (not of Edison origin) in a custom cabinet. *Courtesy of Stan Stanford (Value code: K)*

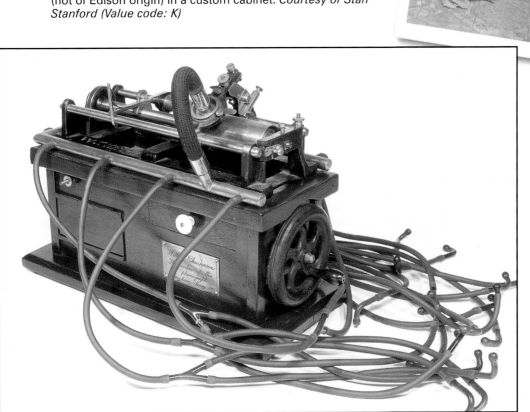

2-20. Edison was the inspiration for this hand-made, hand-driven phonograph. The brass plate reads "William Chas. Tinson, The maker of this phonograph, Longham Surrey 1892." The listening rail suggests a clever entrepreneur who found it cheaper to fashion his own Edison knock-off, with which to give paid demonstrations. *Courtesy of Philippe Le Ray (Value code: VR)*

2-21. The effulgent beauty of this Type "N" Graphophone in its colorful custom carrying case, conjures up depictions of East Indian deities, radiating inner light! Not only are we ready to forgive the Belgian seller for his significant gaffe, Columbia *Phonographe*, we're ready to surrender all to the resplendence of this remarkable 1896 outfit. The long leather strap suggests it might have been employed by an itinerant exhibitor, who toted it from place to place. *(Value code, this particular set-up: VR)*

2-22. This fellow might not look like a go-getter, but observe the appointments of his Type "N" Graphophone, circa 1896. A listening rail with multiple ear-tubes ready for eager customers, *plus* a Columbia flared-bell amplifying horn with support — for those *big* audiences. A fresh box of records, and a table cloth that you can see from across the State. With any luck, they'll be lined up out the door! *Courtesy of the Julien Anton collection (Value code: K)*

2-23. Although the "Hypno-tome" clearly suggests a therapeutic use — the illuminated, faceted sphere that revolved above it might lull a subject into a semi-conscious state — the device was apparently a sham. It was the creation of Muncie, Indiana, fishmonger and real estate entrepreneur Jesse A. Stephens, who most likely used it to attract the attention of potential clients. After all, the medical suffix *tome*, from the Greek, means "to cut or divide" — so, a "Hypnotome" would be an instrument for *removing* someone from a hypnotic state. Chances are, Mr. Stephens wanted to call his contraption a "Hypnophone," but he decided that "Hypnotome" had a more pseudo-medical sound. The machine Stephens adapted was an "Amet," circa 1896: an Edison Phonograph upper works coupled with a spring motor designed by Edward Amet. At the time, Muncie was known as the "Magic City," which at first suggests a connection with hypnotic trances — the "magic" referred to, however, was more likely the phenomenal vigor of the business climate. *Courtesy of Sam Sheena (Value code: VR)*

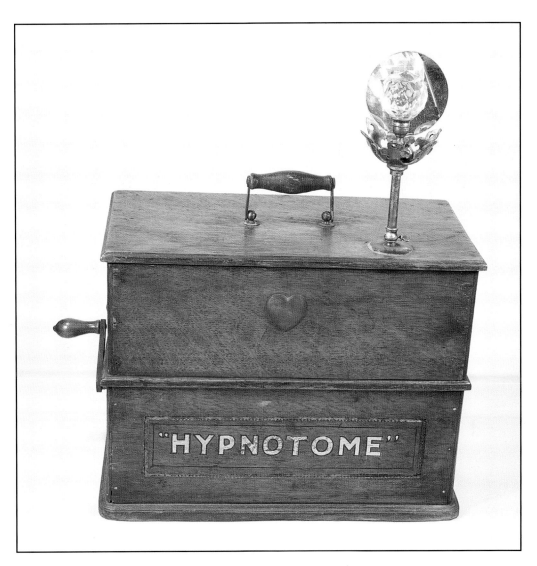

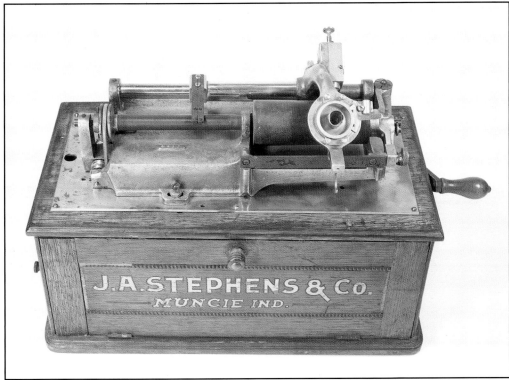

2-24. The "Hypnotome" upper works. *Courtesy of Sam Sheena*

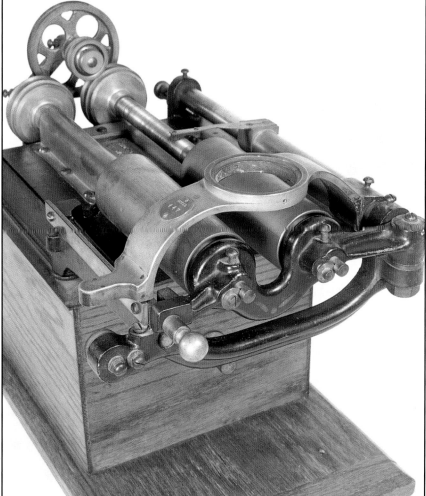

2-25. In the 1890s, the manufacture of cylinder records was an arduous process. At first, artists sang or played "by the round," meaning if six recording machines were running during the performance, the company would have six cylinders to sell. If the company had orders for 100 copies of a title, the artist would need to repeat the performance 17 times. By 1892 Edison had a duplicating process in place, but the product was expensive and of reportedly inconsistent quality. Most small talking machine firms found it cheaper to make their own recordings on blank cylinders supplied by larger companies such a Edison or Columbia. By the mid-1890s, copying machines or pantographs appeared, and suddenly record companies found that they could take six "master recordings" at a single performance from which they could make perhaps 60-80 copies. Quality would deteriorate as the master wore out, and copies were sometimes sold as "second quality" for a few cents less. Exhibitors, however, demanded the loudest records available, and most record firms found it advantageous to advertise *all* their cylinder records as being "Master Records" or "Originals." The unscrupulous merchandising and wholesale piracy of records made the pre-1902 period quite colorful. Naturally, Edison and Columbia guarded their secrets of recording technique and pantograph design like the Crown Jewels. Shown is a Columbia pantograph formerly from the company's archives. *(Value code: VR)*

2-26. A close-up of the Columbia pantograph shows the extent to which it borrowed from the basic Edison design of its upper works. The unit was electrically driven, since reliable and consistent speed was important. Regrettably, the key element of this device — the stylus/cutter arrangement — is missing. Most surprising is the bit of painted decoration on the swing arm between and below the brass mandrels; an aesthetic touch on an otherwise utilitarian instrument.

2-27. The "Graphophon, System Koltzöw" was a specialized hand-driven toy from Germany, circa 1896. It required uniquely-sized cylinder records, which allowed children to record and play through a peculiarly-shaped cardboard horn. Some may notice a similarity to Jean Schoenner of Nuremburg's "Oratiograph" (*The Talking Machine, an Illustrated Compendium*, Fabrizio & Paul, figure 2-99) or Henri Lioret's "Le Babillard" (this book, figure 2-39). Both of these late-1890s toys were based on Koltzöw's device. There is supposition that Schoenner might have made common parts for all three devices. After the turn of the twentieth century, Albert Koltzöw of Berlin was specializing in recording apparatus and records, such as his "Kosmograph" discs. *Courtesy of the Julien Anton collection (Value code: VR)*

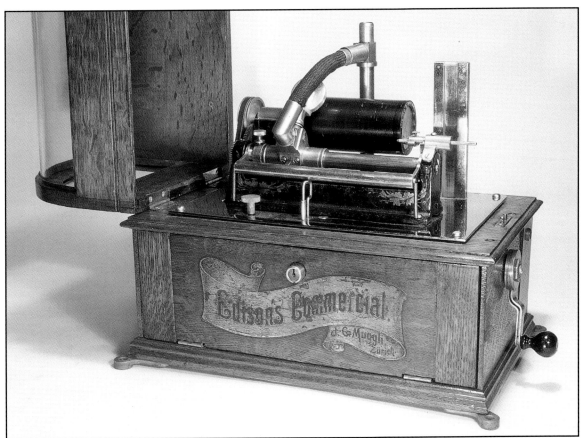

2-28. Columbia was more successful in selling its wares in Europe than rival Edison, but this also gave rise to rampant imitation — counterfeiting on such an enormous scale that it could never be legally pursued. The adaptation of existing Graphophone mechanisms was another by-product of Columbia's success. Many firms had a hand in it, even the virtuous Henri Lioret (see figure 2-38). Commonly, Graphophone serial plates and other identifying marks were removed, covered or effaced. Thus was this rather extreme example of a "faux Edison" created. J.C. Muggli of Zürich, Switzerland, so totally obscured the Columbia origins of this 1897 Type "S" coin-op (serial plate removed, the word "Graphophone" on the upper casting rubbed out, a spurious banner laid directly over the factory decal) that it became an "Edison." "Commercial," in this instance, no doubt refers to the instrument's money-making character. *(Value code: VR)*

2-29. How little it takes to transform a common Columbia Type "B" Graphophone into something extraordinary, and how few such oddities have survived. The Western Talking Machine Company of Omaha (Nebraska) mounted the machine on a cheap, painted wooden plank, covered by an equally inexpensive cardboard lid emblazoned with shameless self-promotion — yet, we're left wishing we had one. *Courtesy of Larry and Myra Karp (Value code: H)*

2-30. These dapper fellows were in all probability aspiring exhibitors, equipped with the Sears, Roebuck, & Company "New Gem" (actually a Type "B" Graphophone) exhibition outfit. The "Talking Machine" signage was probably cut from one of the Sears exhibition posters that were included in the outfit. To our eyes, it might seem an unnecessary label, but these young men may have been touring rural regions where the talking machine was still a novelty known to few. *Courtesy of the collection of Howard Hazelcorn (Value code: K)*

36

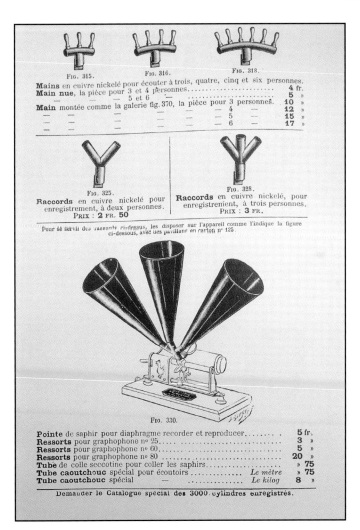

2-31. In France, in 1898, Compagnie Générale de Cinématographes, Phonographes et Pellicules, usually known as Pathé, thoughtfully provided this attachment and three horns for recording a trio. The device was not meant for reproduction. *Courtesy of the Scott and Denise Corbett collection*

2-32. An actual attachment is here shown on a Type "B" Graphophone, ready to make a home recording of a trio. *Courtesy of the Scott and Denise Corbett collection (Value code: VR)*

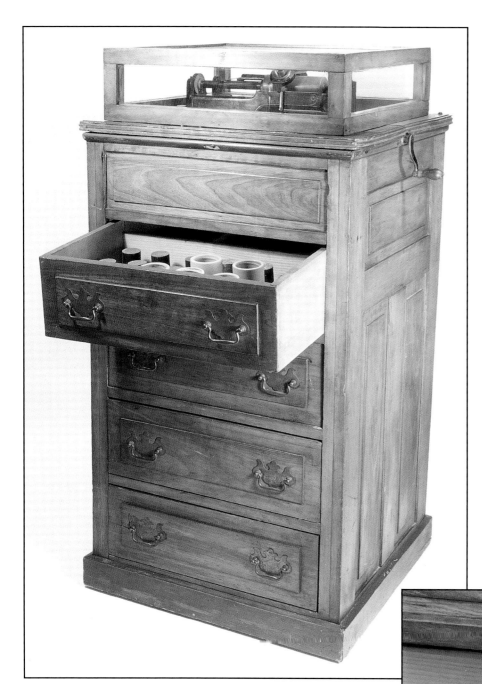

2-33. Hawthorne & Sheble manufactured cabinets to enclose talking machines and records beginning in the 1890s. These apparently inspired a talented craftsman to make one of his own. Perhaps a bit clunky by comparison, this cabinet nevertheless added stature to the Edison "Home" within. The top drawer could accommodate 18 cylinders plus two small compartments for accessories. The other three drawers held 25 cylinders each. Measures 22" wide x 20" deep x 43" high. *Courtesy of the Scott and Denise Corbett collection (Value code: VR)*

2-34. The interior of the top drawer features a flag cut from a title slip of the United States Phonograph Company of Newark, New Jersey (1893-1897). *Courtesy of the Scott and Denise Corbett collection*

2-35. Peter Bacigalupi was one of the West Coast's foremost talking machine entrepreneurs, commencing in the late 1890s. He sold Edisons and Columbias initially, later adding Victors to his stock. This Type "B" Graphophone bears a Bacigalupi stencil, indicating the firm's original address, occupied from 1895 until a fire destroyed the premises in November 1898. *Courtesy of Bob Thomsen (Value code: I)*

2-36. This "two-clip" Edison "Standard" Phonograph from 1899 carries the Bacigalupi decal on its front panel. *Courtesy of Jerry Blais (Value code, with special marking only: G)*

2-37. Beneath the baseboard of the 1899 "Standard," Bacigalupi's distinctive designation was stenciled. Note the change of address to 933 Market, where the firm conducted business until April 1906, when Bacigalupi endured the catastrophic San Francisco earthquake. His spirit was undeterred, as evidenced by his words quoted in the June 1906 issue of *Talking Machine World*: "… I can positively state that $150,000 in coin would not place me where I stood on the 17th day of April, 1906. These places represented the labor of thirty-eight years of my life, but I am not discouraged…" He recovered rapidly, aided in part by a flair for real estate speculation. By May 1907, the *Talking Machine World* could report that Bacigalupi was enjoying "a splendid business." *Courtesy of Jerry Blais*

2-38. In 1899, Henri Lioret realized that he had to expand his market or wither. The instrument shown here, known as the "Spring Driven Eureka," was the French inventor's first effort to move away from his own esoteric, proprietary system into the realm of the "standard" wax cylinders that were becoming increasingly popular throughout the world. To accomplish this, Lioret adapted a conventional Columbia "AT" Graphophone. Stripped of its American wooden cabinet, the Graphophone mechanism was mounted on four robust pillars, and enclosed under a walnut cover (not shown). Note, however, the spoked fly-wheel mounted to the mandrel arbor. This modification was made to keep the speed even, and had been the source of inspiration for the entire project, since Lioret had expressed dissatisfaction with the regulation of contemporary Graphophones. As witnessed in other European machines "adapted" from regular American stock, the ID plate was removed from the Columbia works, and, although it was not done here, the word "Graphophone" was sometimes effaced on the upper casting (see figure 2-28). Lioret offered the machine with ordinary "wax cylinder" accouterments, and/or a characteristic Lioret horn and reproducer supported by a table stand to play special celluloid cylinders that conformed to the "AT" mandrel and specific speed of the mechanism. Shown is a rather imaginative combination of Lioret reproducing arrangements, with wax cylinder in place. *Courtesy Musée de l'Aventure du Son, St. Fargeau (Value code: VR)*

2-39. "Le Babillard" (the chatterbox) was French phonograph pioneer Henri Lioret's turn-of-the-twentieth-century toy, capable of recording and playing back "kid-sized" wax cylinders. The 1899 version, shown here, could also play special pre-recorded celluloid records. The simple, but conventionally designed carriage on the right, to which a combination recorder/reproducer is mounted, was employed for the wax cylinders. On the left, the red horn/reproducer apparatus was used for the celluloid records. Lioret had been producing celluloid cylinders in various sizes since 1895. Notwithstanding the toy shown here, the instruments he manufactured to play his unbreakable records were beautifully rendered and precise — though pricey. Regarding "Le Babillard," there is evidence to suggest that parts of it were made by German toy maker Jean Schoenner, who contributed to two similar phonograph toys. *Courtesy Musée de l'Aventure du Son, St. Fargeau (Value code: VR)*

The Mackie Music Company of Rochester, New York, was an institution closely associated with the phonograph for many years. In 1878, it was the site of some of the earliest exhibitions given with the newly-invented Edison Phonograph. In the late-1890s and early twentieth century, Mackie sold Edison Phonographs and records. Henry S. Mackie's establishment was located at 82 State Street (no longer standing, now the site of a hotel), and comprised an entire building devoted to the sale of musical instruments. Furthermore, the top floor (the fifth, and fortunately accessible by elevator) contained a hall that could be used either for a rehearsal space or for public performances. Mackie's citadel of music, dominated by a large gothic tower, played an important role in Rochester, a city long noted for its connection to musical art.

CONCERT HALL!
State Street.

For a Limited Season, commencing

Monday, May 20, at 10 A.M.

J. S. VALE, - - Manager.

WONDERFUL!
AMUSING!
INSTRUCTIVE!

EDISON'S
Greatest Invention,
THE SPEAKING

PHONOGRAPH

— OR, —

TALKING
MACHINE.

The Phonograph excites the curiosity and admiration of all. It will clearly and accurately repeat whatever may be said into it; declaims, recites prose or poetry, sings and whistles, reproduces the noises of animals and birds, as well as the notes of musical instruments, and is the scientific wonder of the day.

To specially accommodate those who not only desire to hear the Phonograph speak, but also to inspect and examine it closely, several exhibitions of its powers will be given during the day, when those present will be afforded opportunity to test and examine the machine. Those desirous of so doing should make a point to attend the day exhibitions, as at the evening entertainment the privilege of examining the Phonograph can necessarily be only afforded to a very few.

Mr. VALE will fully explain the machine and test its powers at every entertainment, making it reproduce recitations, songs and cornet solos, etc.

Exhibitions at 10 to 12 A.M., 2 to 5 & 7 to 10 P.M.

To prevent the noise of persons coming in late, it is particularly requested that visitors will be in the hall a few minutes before the hours named, so that the exhibitions may commence promptly on time.

Admission, 25 Cents.
Special Rates to Schools. Children under 12, 15 cts.

Daily Union and Advertiser Company's Print.

2-40. In the fall of 1878, the house publication, *Mackie's Bell Treble* (also the name of the firm's brand of pianos), reprinted a newspaper article describing a "Matinee Musicale" at 82 State Street: "At the top of his large building is a hall that is capable of seating about 500 people, with a stage erected at one end of it. Long before four o'clock, the hour announced for the commencement of the entertainment, the hall was packed so closely that it was impossible even to ascend the stairs… On the conclusion of the concert, an invitation was extended by Mr. Mackie to those who took part in the programme, and friends, to partake of a handsome supper spread out at the Temperance Lunch Rooms." It was in this performance space that J.S. Vale, one of the earliest Phonograph exhibitors, held demonstrations of the miraculous device. These presentations began on Monday, May 20, 1878, a mere month after the Edison Speaking Phonograph Company had been founded. Needless to say, this was the first opportunity for everyone attending to apprehend Edison's great invention. Mr. Vale, of New York City, had purchased his apparatus that same month from the Edison Speaking Phonograph Company. His was the fourth entry in the firm's sales book. What Mr. Vale received for the considerable sum if $95.50 was a "Tin Foil" Phonograph manufactured by Patrick & Carter or the Hope Machine Works (the relative value of the machine in 2006 money is about $1850.00, using the Consumer Price Index. More tellingly, if we make a comparison based on the average wages of unskilled laborers in 1878, we arrive at a truly enlightening $13,500.00). 1878 was a fast-paced year for things Phonographic — the line cut used to illustrate this flyer had appeared in *Frank Leslie's Illustrated News* on April 20, 1878. It depicted French soprano Marie Roze singing into a ("Brady" model) "Tin Foil" Phonograph, which event was described as "a marvelous success." A handbill measuring 4 1/2" x 12 1/4". *(Value code: VR)*

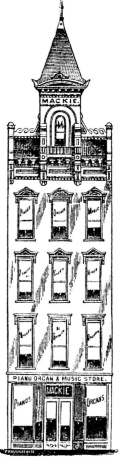

2-41. This line cut depicts Mackie's headquarters on State Street, Rochester, New York.

42

2-42. According to the *Union & Advertiser*, "[Mackie's] building is one of the largest and most complete musical establishments in the country, and reflects great credit on the enterprise of Rochester." It was one of Mr. Mackie's swank clients who no doubt asked that the rather utilitarian Edison "Home" Phonograph of 1898 be elevated, literally and philosophically. The machine was raised on a special base, and then covered with shiny adornment. Rectangular plates on the front and back that might suggest to the viewer places for personalized inscriptions, are in fact just more *bling*. *(Value code: VR)*

2-43. A spirit level, completely unnecessary, but looking suitable scientific, was affixed to the motor plate. This ornamental "Home" was accompanied by a wooden record carrying box, similarly garnished.

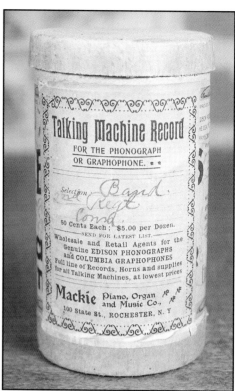

2-44. Mackie's advertising label was pasted over a conventional Edison record container of the period. *(Value code: K)*

2-45. The cabinet of this Edison "Home" Phonograph was evidently too plain for the eye of its original owner. Faux-marquetry decals were applied on all surfaces except the front of the lid, where the original lavish banner was allowed to maintain its prominence. *Courtesy of the collection of Howard Hazelcorn (Value code: VR)*

2-46. The rear of the "Home" was festooned with decoration, including the owner's initials: "L.N." *Courtesy of the collection of Howard Hazelcorn*

2-47. When Albert T. Armstrong's Vitaphone (a putative competitor of Berliner's Gramophone) appeared in 1899, apparently few people outside the talking machine trade took notice. Retailer Kohler & Chase of San Francisco did its best to promote the line, including Joseph Jones's red "Paper Records which are not affected by heat or cold, and are indestructible." This 10" x 11 1/2" flyer is a rare example of Vitaphone promotion. *Courtesy of Don Fenske (Value code: K)*

2-48. The Philadelphia firm of Hawthorne & Sheble was the largest manufacturer of talking machine accessories from 1896 until its bankruptcy in 1909. Its best-selling products were amplifying horns of all types, but during the 1890s and early 1900s the company was well known for its high quality record and machine cabinets. Shown is a variation of its "Nonpareil" cabinet in mahogany ($40.00) equipped for 5" cylinder records and sheltering an 1899 Edison "Concert" Phonograph beneath a glass cover. Five drawers hold 10 cylinders each (50 records total). Measures 37" wide, 19" deep, 51" high, including lid. *(Value code: VR)*

2-49. E. Bouquette, Paris, was a scientific instrument maker who briefly branched into phonographs, circa 1900. The finely finished, lacquered brass components of this machine clearly suggest its origin. *Courtesy of the Julien Anton collection (Value code: VR)*

2-50. The "Victoria" was designed vaguely along the lines of an Edison "Home" Phonograph, but with some inspired touches that would have given old Tom nightmares. Note the mandrel — that's one way of dodging the tapering mandrel patent. This instrument was sold in Europe circa 1900. The horn employed here is paper, with an aluminum bell — it adds very little drag to the carriage, but it's a miracle to have survived. *Courtesy of the Julien Anton collection (Value code: VR)*

2-51. "Great Discoveries of the [nineteenth] Century." The mother encourages her children, "Listen — it's Grandpa that you're hearing!" *Courtesy of the Julien Anton collection (Value code: I)*

CHAPTER THREE

1900–1910

3-1. Edison introduced the "Concert" Phonograph in 1899 to compete with Columbia's "Grand" Graphophones and records, a series introduced the previous year. The "Concert" and "Grand" machines were designed to play 5" diameter cylinder records, considerably larger than those theretofore in use. In short order, both firms adapted coin-operated instruments to embrace the new format. Shown is the "Edison 'M' Concert Coin Slot Phonograph" of 1900. *Courtesy of Sam Sheena (Value code: VR)*

3-2. A close-up showing the mechanism. *Courtesy of Sam Sheena*

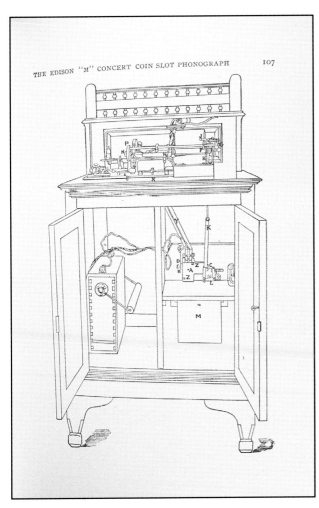

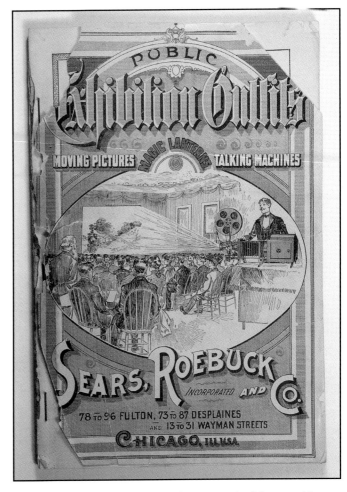

3-3. In *The Phonograph and How to Use It* (shown), published by the National Phonograph Company in 1900, the coin-slot "Concert" was described, "This outfit consists of a Concert Body… mounted on an 'M' Electric motor. It is equipped with an Automatic Reproducer, automatic slot attachment, hearing tube, 24-inch polished brass horn and horn support, together with complete storage battery and cords… As it is actuated by electricity, no winding is necessary. Runs continuously and without noise. Usually operated by a nickel… As a novelty, the Edison Concert Coin Slot Phonograph is the most attractive and wonderful musical and talking machine ever put before the public."

3-4. In the late 1890s, moving pictures and talking machines were all the rage. Each month, it seemed, a new projector was on the market, and dealers clamored for films and sound recordings that would appeal to the public. Sears, Roebuck, & Company foresaw a golden opportunity for exhibitors to cash in on the public's insatiable hunger for entertainment, and for a few years around 1900 the company offered special catalogs of "Public Exhibition Outfits." This example dates from late 1900. *Courtesy of Bill Boruff*

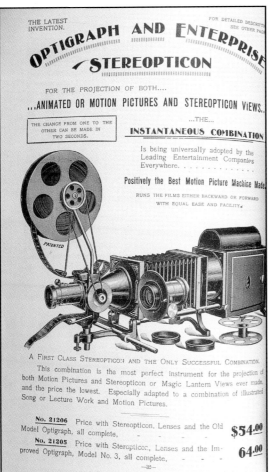

3-5. In this catalogue Sears dealt exclusively with the Optigraph Motion Picture Machine coupled with a stereopticon, both manufactured by the Enterprise Optical Manufacturing Company, Incorporated, located in Chicago. *Courtesy of Bill Boruff*

The Graphophone
or Talking Machine

THE PERFECT MUSICAL AND TALKING MACHINE.

TRIED, TESTED AND GUARANTEED.

PUBLIC ENTERTAINER OF UNRIVALED MERIT AND A MINT OF MONEY FOR THE EXHIBITOR.

USED for public exhibition work, our entertainment outfits described elsewhere, will pay for themselves in from one to three nights and by reason of their simplicity, the ease with which they can be operated, the attractiveness of the programme presented, they are the best and most popular form of outfit for parties desirous of embarking in exhibition work on a comparatively limited capital.

=== AS A MONEY-MAKER ===

The Graphophone Has No Equal.

— 93 —

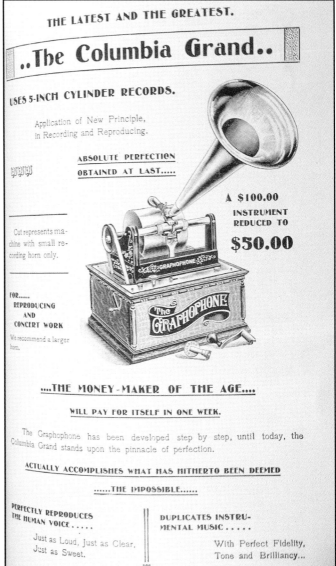

THE LATEST AND THE GREATEST.

..The Columbia Grand..

USES 5-INCH CYLINDER RECORDS.

Application of New Principle, in Recording and Reproducing.

ABSOLUTE PERFECTION OBTAINED AT LAST.....

Cut represents machine with small recording horn only.

FOR...... REPRODUCING AND CONCERT WORK

We recommend a larger horn.

A $100.00 INSTRUMENT REDUCED TO **$50.00**

....THE MONEY-MAKER OF THE AGE....

WILL PAY FOR ITSELF IN ONE WEEK.

The Graphophone has been developed step by step, until today, the Columbia Grand stands upon the pinnacle of perfection.

ACTUALLY ACCOMPLISHES WHAT HAS HITHERTO BEEN DEEMED

......THE IMPOSSIBLE......

PERFECTLY REPRODUCES THE HUMAN VOICE.....	DUPLICATES INSTRUMENTAL MUSIC.....
Just as Loud, Just as Clear, Just as Sweet.	With Perfect Fidelity, Tone and Brilliancy...

— 101 —

3-6. For the talking machine portion of an exhibitor's presentation, Sears recommended the various models of Graphophone, "A Public Entertainer of Unrivaled Merit and A Mint of Money for the Exhibitor." *Courtesy of Bill Borutt*

3-7. For the exhibitor who wanted a first-class Exhibition Outfit, the "Columbia Grand" (Type "AG") was top of the Sears line. *Courtesy of Bill Borutt*

3-8. Sears formed a subsidiary firm called the Entertainment Supply Company to handle the anticipated hordes of enterprising young men purchasing exhibition equipment. Note the mysterious "Grapho-Ampliphone" terminology.

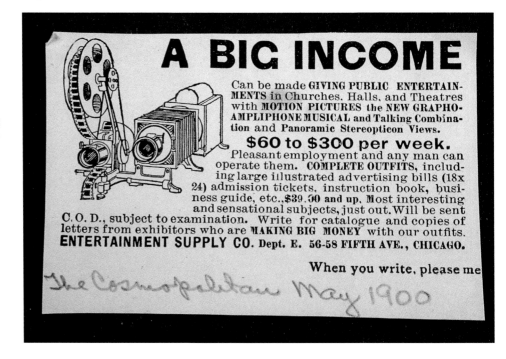

A BIG INCOME

Can be made GIVING PUBLIC ENTERTAINMENTS in Churches, Halls, and Theatres with MOTION PICTURES the NEW GRAPHO-AMPLIPHONE MUSICAL and Talking Combination and Panoramic Stereopticon Views. **$60 to $300 per week.** Pleasant employment and any man can operate them. COMPLETE OUTFITS, including large illustrated advertising bills (18x 24) admission tickets, instruction book, business guide, etc., $39.50 and up. Most interesting and sensational subjects, just out. Will be sent C. O. D., subject to examination. Write for catalogue and copies of letters from exhibitors who are MAKING BIG MONEY with our outfits. ENTERTAINMENT SUPPLY CO. Dept. E. 56-58 FIFTH AVE., CHICAGO.

When you write, please me

The Cosmopolitan May 1900

3-9. The "Grapho-Ampliphone" apparatus was featured in this exhibition poster, whose origin is clearly Sears, Roebuck & Company, although it lacks the usual company attribution at the bottom. *(Value code: J)*

3-11. The connection between the "Grapho-Ampliphone" (simply a Graphophone equipped with this immense horn), the Enterprise Optical Manufacturing Company, Incorporated, and Sears, Roebuck, & Company is represented by this artifact. Since larger cylinder Graphophones such as the "Columbia Grand" were originally supplied with only a small horn, it was natural for Enterprise to offer an appropriately large horn for the Sears outfits that featured Enterprise film projection equipment. *Courtesy of the Gfell family collection (Value code: VR)*

3-10. The mystery of the "Grapho-Ampliphone" was solved by the discovery of this 55" long aluminum horn with 20" diameter bell. See next illustration for a close-up of the horn's markings. *Courtesy of the Gfell family collection (Value code: VR)*

3-12. The Entertainment Supply Company also offered the same standard Columbia cylinders that were seen in the Sears catalogs, but with special labels for the exhibition trade. *Courtesy of the Babcock House Museum and John H. Perschbacher (Value code: VR)*

3-13. Another view of the special cylinder record labeling. *Courtesy of the Babcock House Museum and John H. Perschbacher*

3-14. You're looking at the only surviving example of an Automatic Lioretgraph. In 1900, France hosted a World's Fair, and 300 such machines were installed on the grounds by Henri Lioret. By all reports, the coin-operated mechanisms performed successfully, yet Lioret was still trying to sell off his stock of them in 1903. Evidently, other exhibitors were not convinced of their merits. Three variations were offered, of which this is the "loud speaking" type (that is, it used a horn rather than ear tubes). *Courtesy of Collection Fabrice Catinot, Dijon (Value code: VR)*

3-15. The automatic mechanism was based on existing Lioret technology. It was driven by weight power, like a clock. Dropping a 10 centime coin in the slot released a stop on the crank handle, which could then be rotated, raising the weight and triggering the record to play. Of course, Lioret's celluloid cylinders were perfectly suited to the frequent use of a coin-op, and he created a special series of three minutes duration, less playing time than many of the records he sold to the public, but better for turnover. *Courtesy of Collection Fabrice Catinot, Dijon*

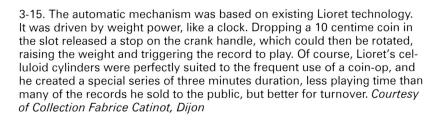

3-16. By the end of 1900, Henri Lioret had all but given up on efforts to sell his superior (up to four-minutes in playing time, unbreakable) but expensive celluloid cylinders. His "Eureka" records cost approximately three-and-a-half to five-and-a-half times the price of the wax competition. It's ironic (and a rather tragic tribute to the French inventor) that M. Lioret was abandoning the plastic record business just as others were tentatively embarking upon it (beginning with Thomas Lambert in the United States). At the close of 1900 Lioret joined in a venture with fellow talking machine entrepreneur Louis Lamazière to produce an instrument that could play the 5" diameter "Grand" and "Concert" cylinders that had recently been gaining popularity in Europe. It was agreed the machine would be offered separately by both men. Lioret marketed it under the name "L'Eclatant" ("ringing" — as in "loud"). When sold by Lamazière, the machine, equipped with a Bettini reproducer, carried a brass plate on the front, as seen here, stamped, "Louis Lamazière Constructeur, Bte. SGDG [patented], 99 Fg. du Temple Paris." The design inspiration was clearly an Edison "Concert." *Courtesy Musée de l'Aventure du Son, St. Fargeau (Value code: VR)*

3-17. Société Française des Phonographes "Réal" of Levallois sur Seine was responsible for this incredible feat of engineering! A wistful combination of ingredients, whipped to a fetching froth. What does the name imply? The root *"réal"* is used in French words such as *"réaliste"* (realistic), which could be an inspiration for the nomenclature. Furthermore, there is the usage observed in the name of the city of Montréal, where *"réal"* means "royal" (translating to "Mount Royal" in English). In this same context, the *real* (*réal* in French) was a unit of Spanish currency for centuries, meaning "the royal coinage." The "Réal" talking machine was one of many fascinating European design solutions for playing 5" diameter "Concert" or "Grand" records. *Courtesy of the Julien Anton collection (Value code: VR)*

3-18. Czempin & Krug, talking machine manufacturers of Berlin, Germany, produced this "Ideal" phonograph. Standing on its four legs, it recalls a machine out of an animated cartoon — all that's needed to complete the image is a smile and a set of eyes. This instrument was advertised in the August 15, 1900 *Phonographische Zeitschrift* trade journal. *(Value code: VR)*

3-19. An embossed metal advertising sign, 12" x 16", from 1900 (the same illustration, by artist Lorant-Heilbron, may be seen on the cover of the July 1900 Pathé catalog). Note that in this early period of Pathé activity phonographs were always mentioned in conjunction with "film projectors and other precision instruments." An angel, bearing an Orphean harp, elevates a Pathé cylinder machine (no disc instruments yet) surmounted by an excited rooster. Dramatic! — yet, the most interesting aspect of the piece can be found below (see next illustration). *Courtesy of Jalal and Charlotte Aro (Value code: VR)*

3-20. At the lower right of the Pathé metal sign, a throng of people clamber up a slope leading to phonographic delights, and clamor for material fulfillment. Although some men are present, it is women who are making the most vigorous effort to reach the talking machine goodies, as if having a peculiar little wooden box containing greasy clockwork in the corner of the *salle de séjour* were essential to their happiness. One is reminded of the female penchant for shopping, and the retailer's understanding of it. It was not by chance that the fictitious department store, from which Zola's 1883 novel took its name, was called Au Bonheur des Dames (Ladies' Delight). The fin de siècle world was one where even ordinary folk could at last enjoy the gratification of buying things. In Great Britain, Symonds' London Stores published a monthly journal, to which customers could subscribe for a yearly fee of one shilling, entitled Pleasure. "Including an important department devoted to… Phonographs, Gramophones, Graphophones and other Talking Machines…" *Courtesy of Jalal and Charlotte Aro*

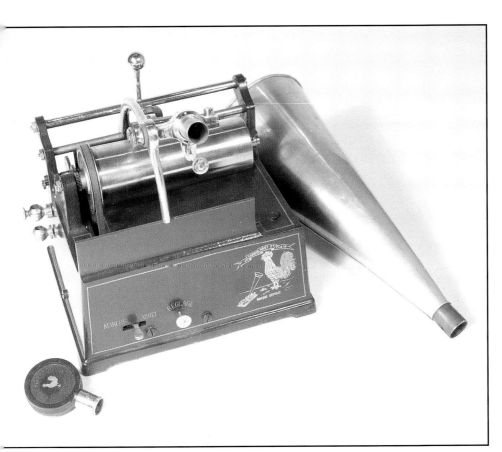

3-21. Pathé copied the design of the 1899 Edison "Gem" Phonograph, to produce one of its most popular cylinder talking machines. The instrument was introduced in 1900 under a dual title. The top heading, "Graphophone No. 20" (or "21," depending on the catalog), has been largely forgotten in favor of the catchier sub-heading, "Le Gaulois" (the Frenchman). We hope the purists among you can forgive Pathé for referring to an imitation Edison product as a Graphophone, instead of a Phonograph. To support the usage, one can point to the fact that the carriage and reproducer of "Le Gaulois" were appropriated from Graphophone technology, but in truth Pathé used Phonograph and Graphophone interchangeably — the high-stakes struggle to establish the primacy of one term over the other took place mainly in the United States. The 1900 catalog mentions that "Le Gaulois" could be ordered in red, black or gray, with the customer's choice respected wherever possible. "Overpainted" examples have been seen (for instance, blue over red) suggesting Pathé went to considerable length to accommodate its patrons. Over the years the colors came to include black, green, red, blue, gray and silver. Seen here is a previously unknown color, what the French would call *marron* (chestnut brown). The English word "maroon" is closely related, which calls to mind Edison's Model "D," or "Maroon" "Gem" Phonograph of 1909. *(Value code: VR)*

3-22. This Pathé outfit included a fairly modest instrument, the redoubtable "Coq," but in such an elegant context that it looks fit for a king. It was referred to as the "Bôite-Nécessaire de Graphophone 'Coq'" ("Cock" Graphophone kit). *Courtesy of Garry James (Value code: G)*

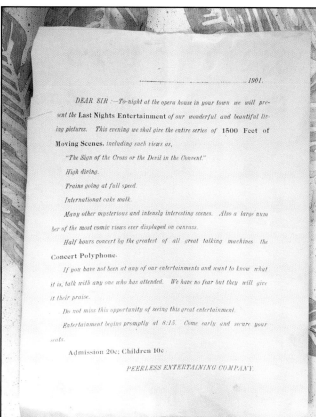

3-23. Traveling exhibitors were common at the turn of the twentieth century. Traces of their activities continue to surface, such as this flyer from 1901, inviting the recipient to view "1500 Feet" of new-fangled motion pictures and a "half hours concert by the greatest of all great talking machines[,] the Concert Polyphone." The "Peerless Entertaining Company," like many other such enterprises, was too small to make much of a dent in history. *(Value code: VR)*

3-24. In addition to the flyer in the previous illustration, this cylinder record traveling case with its 24 Edison Concert 5" cylinders somehow survived for over a century. This unusually-designed case was manufactured by the Automatic Slot-Handle Box Company, of Chicago, Illinois. These records would have provided the "Peerless Entertaining Company" with approximately 48 minutes of diversion for its clientele. See next illustration for one more artifact from this itinerant business. *(Value code: VR)*

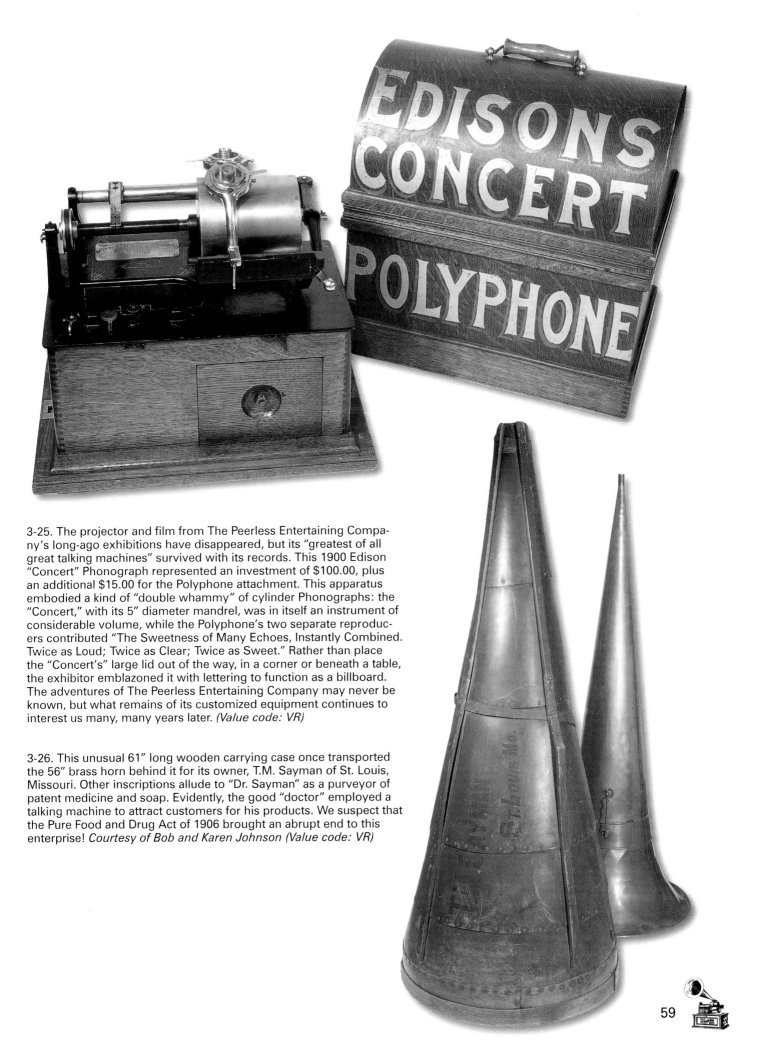

3-25. The projector and film from The Peerless Entertaining Company's long-ago exhibitions have disappeared, but its "greatest of all great talking machines" survived with its records. This 1900 Edison "Concert" Phonograph represented an investment of $100.00, plus an additional $15.00 for the Polyphone attachment. This apparatus embodied a kind of "double whammy" of cylinder Phonographs: the "Concert," with its 5" diameter mandrel, was in itself an instrument of considerable volume, while the Polyphone's two separate reproducers contributed "The Sweetness of Many Echoes, Instantly Combined. Twice as Loud; Twice as Clear; Twice as Sweet." Rather than place the "Concert's" large lid out of the way, in a corner or beneath a table, the exhibitor emblazoned it with lettering to function as a billboard. The adventures of The Peerless Entertaining Company may never be known, but what remains of its customized equipment continues to interest us many, many years later. (Value code: VR)

3-26. This unusual 61" long wooden carrying case once transported the 56" brass horn behind it for its owner, T.M. Sayman of St. Louis, Missouri. Other inscriptions allude to "Dr. Sayman" as a purveyor of patent medicine and soap. Evidently, the good "doctor" employed a talking machine to attract customers for his products. We suspect that the Pure Food and Drug Act of 1906 brought an abrupt end to this enterprise! *Courtesy of Bob and Karen Johnson (Value code: VR)*

3-27. The Victor Talking Machine Company briefly offered an ornately carved oddity in late-1901-1902 called the "Monarch Deluxe." This $60.00 creation employed the same mechanics as the current top-of-the-line "Monarch" but with over-the-top cabinetry. Very little was seen in company advertising relating to this model, but, in late-1901, Victor crowed about the gold medal it received at the Pan-American Exposition in Buffalo, and included an illustration of the short-lived "Monarch Deluxe." *Courtesy of Ernest Carl Allen (Value code: VR)*

3-28. One of the few known surviving examples of the Victor "Monarch Deluxe." This one features an all-brass horn. *Courtesy of Ernest Carl Allen*

3-29. A close-up of the "Monarch Deluxe's" distinguishing feature: the heavily carved cabinet. Although this model has taken on mythic proportions among modern collectors, evidently few turn-of-the-twentieth-century customers found the woodwork (or the lofty price) appealing, making this arguably the rarest of Victor's external-horn models. *Courtesy of Ernest Carl Allen*

3-30. The 1901 Pan-American Exhibition in Buffalo, New York attracted many firms that recognized its potential for public exposure. No doubt Eldridge Johnson's new advertising virtuoso from Chicago, Leon Douglass, was enthusiastic about showing Victor Talking Machines and Victor Records to the throngs crowding the exhibits. The fledgling company (not yet incorporated as Victor) went a step further in a clever publicity stunt, and gave away 3" diameter souvenir disc records. Recipients who did not already own a Victor were encouraged to "Ask your nearest Music Dealer to let you hear what this Record has to say about the Victor Talking Machine." It's interesting to note that this disc was pressed from celluloid, which Johnson's predecessor, Emile Berliner, had flirted with, but had abandoned in favor of a shellac compound. *Courtesy of Larry and Myra Karp (Value code: VR)*

3-31. The reverse of the 3" Victor souvenir disc carried an even more impressive message. The "March King" himself, John Phillip Sousa, gave his blessing to Johnson's enterprise. Later, in the September 1906 issue of *Appleton's Magazine*, Sousa reversed his stance, and railed against what he considered "canned music." By that time, however, other celebrity endorsements for the Victor were easily secured — many artists having made their fortunes because of their Victor recordings. Despite his printed tirade on "The Menace of Mechanical Music," Sousa, too, would garner considerable fame from his recording contracts. *Courtesy of Larry and Myra Karp*

3-32. Edison's first model of the "Gem" Phonograph (1899) had a significant impact on Europe. Pathé rapidly co-opted the design for its "Le Gaulois" series. In Germany, Allgemeine Phonographen-Gesellschaft m.b.H. (General Phonograph Company, Limited) of Krefeld, a firm commonly known as "Krefelder," copied the "Gem" in 1900 as seen here. In this version, the mechanism was provided with a metal cover, sometimes marked "Mignon B." Note the carriage, equipped with a "floating" reproducer, combining Graphophone with Edison technology. It was a great idea — the cheesy little reproducer supplied with the first Edison "Gem" was marginal. In the context of European imitation, there were no legal or emotional reasons not to

mix the best ideas of the rival American firms. The impunity with which the Edison "Gem" was imitated provoked the following comments in the 1905 catalog of phonographs issued by the Britain's Symonds' London Stores: "It [Edison's 'Gem'] is being counterfeited extensively and placed upon the market… unless you get the genuine, you would do better not to buy at all… If you buy an imitation Edison Phonograph, you not only get an inferior makeshift, but it is liable to make legal trouble for you, because Mr. Edison has already brought legal actions against several who have been handling bogus goods…" The scope of such prosecutions proved to be exhaustingly vast, and Edison preferred hectoring his domestic dealers for the tiniest infraction. *Courtesy of Jürgen Bischoff (Value code: H)*

3-33. The 1902 version of the Krefelder "Gem" look-alike was painted a muted shade of green. For those familiar with toy trains, it's the same color as a Lionel "33" engine — what should we call it? Extreme olive? Please note that the spun metal horns popular in Europe during the first decade of the twentieth century were made of brass, as well as (more commonly) aluminum. *Courtesy of Jürgen Bischoff (Value code: H)*

3-34. Berliner's "Trade Mark" Gramophone, as manufactured by Eldridge Johnson's machine shop starting in 1897, had an enduring worldwide legacy. Johnson continued to make them after he took over the reins of the Gramophone interest in the United States, and the instrument remained popular in Europe during the first years of the twentieth century, when this coin-op version was offered by Compagnie Française du Gramophone. There had been only one, unsuccessful attempt to adapt "Trade Mark" Gramophones to coin operation in the United States. It is interesting to note that Victor Talking Machines, which evolved from the Gramophone in America, were never customized for slot use until the late 1920s. In Great Britain and Europe, however, there were two effective coin-op modifications of the "Trade Mark." The version seen here employed the larger horn of the two. *Courtesy Musée de l'Aventure du Son, St. Fargeau (Value code: VR)*

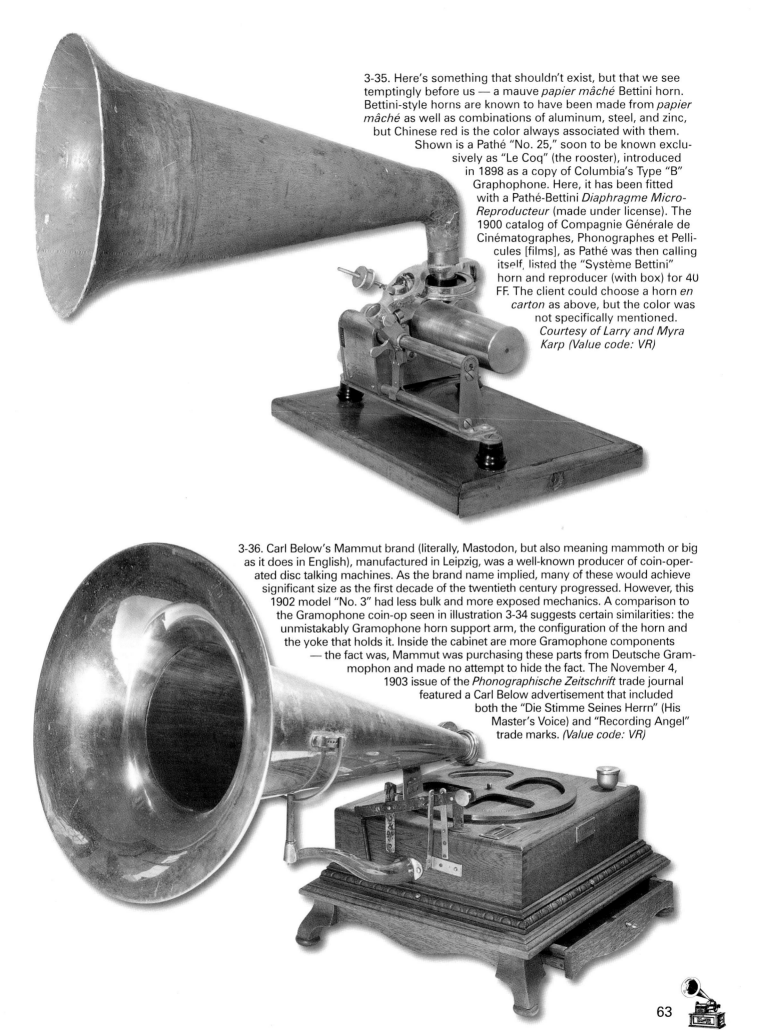

3-35. Here's something that shouldn't exist, but that we see temptingly before us — a mauve *papier mâché* Bettini horn. Bettini-style horns are known to have been made from *papier mâché* as well as combinations of aluminum, steel, and zinc, but Chinese red is the color always associated with them. Shown is a Pathé "No. 25," soon to be known exclusively as "Le Coq" (the rooster), introduced in 1898 as a copy of Columbia's Type "B" Graphophone. Here, it has been fitted with a Pathé-Bettini *Diaphragme Micro-Reproducteur* (made under license). The 1900 catalog of Compagnie Générale de Cinématographes, Phonographes et Pellicules [films], as Pathé was then calling itself, listed the "Système Bettini" horn and reproducer (with box) for 40 FF. The client could choose a horn *en carton* as above, but the color was not specifically mentioned. *Courtesy of Larry and Myra Karp (Value code: VR)*

3-36. Carl Below's Mammut brand (literally, Mastodon, but also meaning mammoth or big as it does in English), manufactured in Leipzig, was a well-known producer of coin-operated disc talking machines. As the brand name implied, many of these would achieve significant size as the first decade of the twentieth century progressed. However, this 1902 model "No. 3" had less bulk and more exposed mechanics. A comparison to the Gramophone coin-op seen in illustration 3-34 suggests certain similarities: the unmistakably Gramophone horn support arm, the configuration of the horn and the yoke that holds it. Inside the cabinet are more Gramophone components — the fact was, Mammut was purchasing these parts from Deutsche Grammophon and made no attempt to hide the fact. The November 4, 1903 issue of the *Phonographische Zeitschrift* trade journal featured a Carl Below advertisement that included both the "Die Stimme Seines Herrn" (His Master's Voice) and "Recording Angel" trade marks. *(Value code: VR)*

3-37. In Europe, imitation was not only the sincerest form of flattery — it was business as usual! American talking machines were brazenly copied with impunity, since patent litigation was unlikely or impossible. Europeans sometimes borrowed from themselves, however. This Eisemann "Matador" is a German machine suspiciously similar to the French Idéal (which was also available in plain brass finish). Even the French instrument's "Le Cahit" reproducer, its customary storage box, and the U-shaped weight for increasing volume have been shamelessly faked. Ernst Eisemann & Co., of Stuttgart, advertised "Phonographen und Automaten [phonographs and coin-ops], records unter Garantie fur Original-Aufnahmen [guaranteed original recordings]." *Courtesy of Jürgen Bischoff (Value code: VR)*

3-38. Leipzig in Germany was an industrial center famous for, among other things, the production of mechanical music devices and talking machines. Lipsia refers to the ancient name of the city. The "block" construction of the motor housing, and the brake lever are typical of Excelsior. However, the "bridge" which spans the length of the mandrel and prevents the horn assembly from falling is unusual. *Courtesy of Jürgen Bischoff (Value code: H)*.

3-39. Some "Gaulois" instruments were manufactured with painted wooden bodies capped by a casting that supported the both motor and upper mechanism. It seems "logical" that this version would have been the first, soon found to be impractical and replaced by a sturdier all-metal model. A tempting conclusion, but not supported by the evidence. An examination of the type of carriage arm employed in the wooden version betrays a style more appropriate to 1903. It's likely, therefore, that the use of wood was an effort to economize. Seen here, is a wooden model supporting a glass horn with "metallic" (that is, iridescent) finish. These horns have a pale pink or yellow (shown) tint. *Courtesy of Garry James (Value code, including glass horn: D)*

3-40. This German Excelsior machine incorporates elements not commonly seen on cylinder phonographs of similar age or size, and raises some questions about the purpose for which it was originally intended. Was it a home entertainment, or an office machine? Note the two limits of the speed scale — 160 rpm, the "standardized" speed of pre-recorded cylinders (after 1902), and 100 rpm, which would have been best suited to recording dictation in a business setting. Furthermore, there is the precise scale, with pointer, that corresponds to the available space on the record. There is also a belt guard, a highly unusual feature. Since this instrument is so out of the ordinary, it is worth noting the "button" on the top front of the cabinet

that releases the hinged "shelf" on which the mechanical works are mounted, allowing access to the motor. Most modestly-sized European machines of the period did not include this convenient feature. There are many reasons to associate it with the practical concerns of an office machine — yet, it has a mounting post on the left side of the works for attaching a horn support, more commonly connected with home use. *Courtesy of Jean-Paul Agnard (Value code: H)*

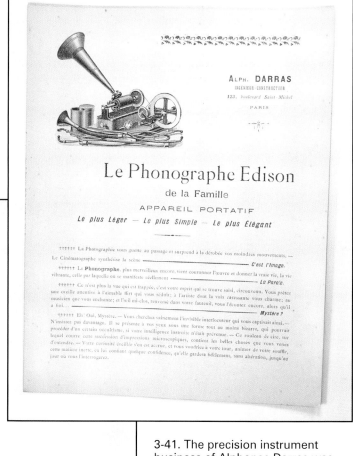

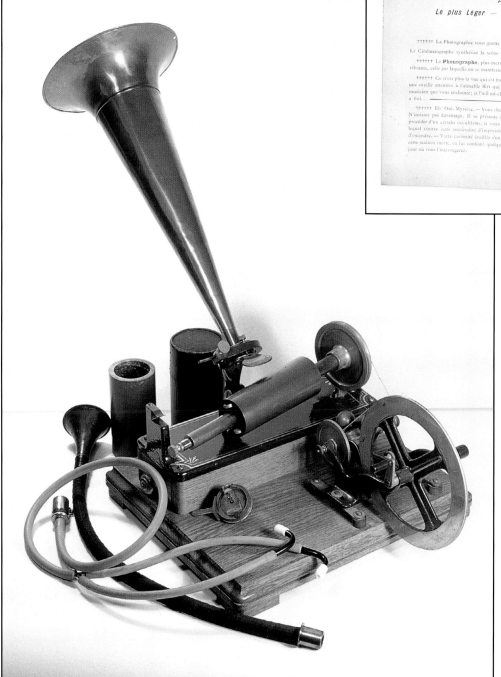

3-41. The precision instrument business of Alphonse Darras was founded in 1866. Over the course of the next 50 years, the company made (in addition to the phonograph shown here) telephone and telegraph apparatus, speedometers, taximeters, controls and relays, and arithmometers (mechanical calculating machines). At first glance, everything about this imaginative hand-driven mechanism looks unique — but, note the conventional Edison recorder and Model "C" reproducer from the 1902 era. *Courtesy of Jalal and Charlotte Aro (Value code: VR)*

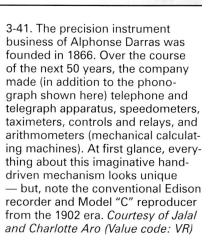

3-42. Darras made no mystery of the fact that he had incorporated an ordinary Edison reproducer and recorder into his device. The instructions give Edison credit — though the American inventor might not have taken kindly to the homage. Edison fought with Edison Bell in England for years over the semi-legal appropriation of his Phonograph designs. M. Darras did not even have "semi-legality" on his side. But Darras was far away, and Edison had individuals closer to home who dared to tamper with his reproducers — such as Edwin Mobley, upon whom Edison heaped punishing prosecution. *Courtesy of Jalal and Charlotte Aro (Value code: VR)*

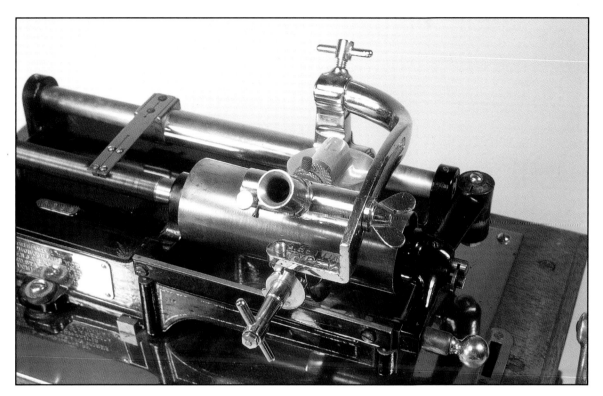

3-43. Of the legendary apparatus purveyed by Gianni Bettini, much has been written. His prescient understanding of the importance of diaphragm size, material, and tension, as well as a non-constricted sound pathway, resulted in the best sound reproduction of the 1890s and early 1900s. Customers desiring Bettini equipment to enhance their Edison Phonographs or Columbia Graphophones were expected to pay a premium for it. Most curious, then, is this Bettini Type "V" arm described in a 1902 Bettini catalog as follows: "For those who wish to use the new style Columbia Recorder on Edison machines, in connection with Bettini arms, we make a special arm V, price $6.00." As Bettini's expensive, specialized equipment lost favor, he compromised his technology to reflect his rivals' inferior paradigms. By 1902 Bettini had moved his headquarters to Paris, where eventually he bestowed his trademark upon ordinary cylinder talking machines with barely a trace of the innovative components he had pioneered. *Courtesy of Douglas Defeis (Value code: VR)*

3-44. Was Gianni Bettini stealing from Edison's "Gem," Pathé's "Gaulois," or Krefelder's "Mignon B" when he produced this Bettini "No. 3" machine? In the incestuous world of the talking machine industry, anything worked if you could get away with it. The reproducer is a "Le Cahit," more commonly associated with the "Idéal" brand. *Courtesy of Richard and Jill Pope (Value code: VR)*

3-45. Gianni Bettini, inventor and entrepreneur, is best known for the "Micro-Reproducers" he made in the United States and France to adapt conventional Edison and Columbia cylinder talking machines. His brief foray into the disc business is less well-documented. In 1903 and 1904 (until his enterprise went bankrupt in June), Bettini, then located in France, attempted to break into the disc market with a small line of instruments and records. The fact that he did not succeed has made Bettini disc artifacts extremely rare today. Shown is a "No. 20," the elegant cabinet of which reflected the style of Bettini's cylinder model "No. 4." *Courtesy of Philippe Le Ray (Value code: VR)*

3-46. Bettini was proud of the name he'd made for himself in the phonograph industry, and his signature was characteristically emblazoned on his products. Even the imaginatively designed soundbox (looking like a cross between a Zonophone "V Concert" and miniature exercise equipment) was engraved with his autograph. *Courtesy of Jalal and Charlotte Aro (Value code: VR)*

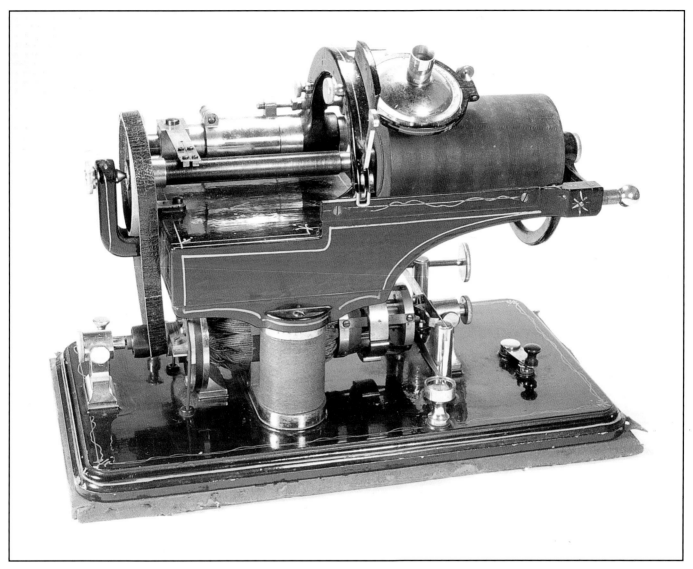

3-47. This unknown European instrument was clearly inspired by Edison's Phonographs (such as the "Home" or "Triumph"). Edison design elements are present, but the electric motor, instead of looking utilitarian and being hidden within a cabinet, has been made shiny and bright and eye-catching. The machine was adaptable to both standard and "Inter" cylinders, which suggests a date of approximately 1903. *Courtesy of Jalal and Charlotte Aro (Value code: VR)*

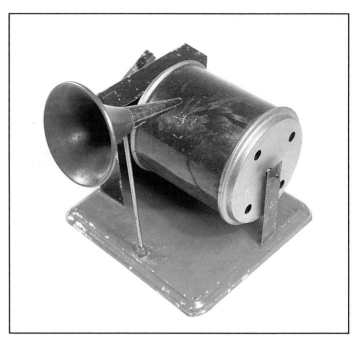

3-48. This spring-driven toy phonograph on a 4" x 4" base resembles the products of German manufacturers such as Bing. Whatever its origin, the fanciful likeness represents a form of entertainment that had become commonplace during the first decade of the twentieth century. *Courtesy of the Tom and Sandi McCarthy collection (Value code: I)*

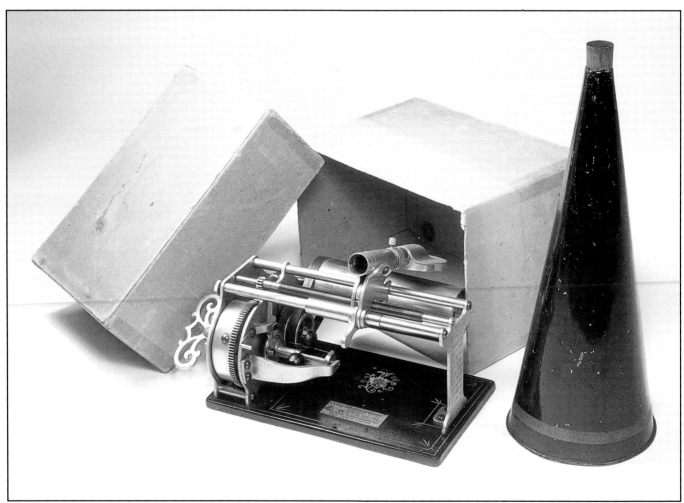

3-49. It's quite unusual to discover a century-old talking machine in its original shipping container. This Type "Q" Graphophone of 1903 was evidently never used, and looks just as it did when it left the factory over a century ago. The horn, deprived of a box, suffered a few minor scratches, but one isn't likely to see a nicer condition "Q" than this. *Courtesy of the Johnson Victrola Museum (Value code, this particular one in box: VR)*

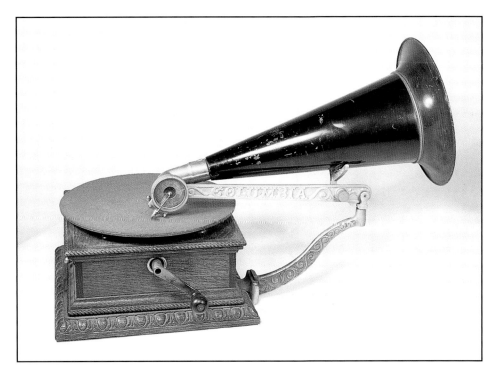

3-50. In the fall of 1901, the Columbia Phonograph Company introduced its line of Disc Graphophones, consisting of the $30.00 "AH" and the $20.00 "AJ" (shown). Both of these models underwent rapid changes during the next two years, particularly the "AJ." The very first version had a vertically-mounted winding crank (in the manner of the "Trade Mark" Berrliner Gramophone). This example, from 1903, retained the early-style cabinet, but with a reconfigured winder in the side position. A later version simplified the cabinet molding. *Courtesy of Bert Gowans (Value code: G)*

3-51. The previous example with its early-style cabinet removed shows the single spring motor mounted from above. Earlier "AJ" models had their motors attached to the baseboard. The 16" black japanned horn was standard equipment on several Columbia models. *Courtesy of Bert Gowans*

3-52. Stollwerck was a German chocolate manufacturer that came up with a thoroughly "up-to-date" marketing scheme: selling a toy phonograph that played little (3 1/8") disc records pressed of chocolate. When a child had tired of a particular record, he could eat it — one of the earliest examples of recycling! In fact, Stollwerck also pressed records from wax ("karbin"), not nearly as appealing to the palate. Stollwerck devices appeared in several variations. In France, they were distributed by the firm of Kratz-Boussac, which sold them under the "Eureka" brand. The all-metal, lithographed 1903 version is shown here with its original box, looking utterly untouched by the hand of time. *Courtesy of René Rondeau (Value code: D)*

3-53. This wooden-cased Stollwerck, with grain-painted metal horn, was released in 1904. As in the previous model, the turntable was driven by a diminutive spring motor. A record carton, from which one disc has been withdrawn, can be seen at the rear. *Courtesy of Jalal and Charlotte Aro (Value code: VR)*

3-54. Even the tiny reproducer was grain-painted. With this machine, the company also introduced a series of 4 3/4" discs made of composition material. Unfortunately, the public did not embrace the notion of a tiny, elegant talking machine, and Stollwerck's brief experiment with music machines ended rapidly. *Courtesy of Jalal and Charlotte Aro*

3-55. A promotional card for meat extract. There must be at least one monkey's uncle among this group of phonograph enthusiasts. *Courtesy of René Rondeau (Value code: K)*

3-57. This comfortable room was in the New Haven, Connecticut home of Albert William Honywill, District Superintendent for the New York, New Haven, and Hartford Railroad. Prominent in Mr. Honywill's parlor was a Victor "Monarch Special" sitting atop a Herzog record cabinet — and one might wonder if he preferred recorded entertainment to strumming on the old banjo. *Courtesy of Paul W. Horgan (Value code: K)*

3-56. The Victor "MS" was one of the company's top instruments from 1902-1906. At $45.00, most customers could ask nothing more of a disc talking machine. However, for those who wanted maximum sound for exhibitions or outdoor entertainment, Victor stood ready to oblige. A floor stand and extra-long traveling arm were available for $4.00, which allowed the use of horns up to 56" in length. You can imagine the kind of inertia this horn/arm/soundbox configuration would have, and it had to be dragged across the record by the groove. Woe to the 78 that was played on this contraption (available briefly in 1901-1902)! *(Value code: VR)*

3-58. An Edison "Home" Model "A" featuring a mahogany cabinet (a $10.00 option) and nickel-plated upper works (a $25.00 option). These embellishments elevated the 1904 price of this example to $65.00. *Courtesy of Bob and Karen Johnson (Value code: E)*

3-59. American expatriate and entrepreneur F.M. Prescott's International Zonophone Company excelled at unusual and eye-catching designs, which reached their zenith after Prescott had left the firm during 1903. Zonophones in the United States were known for graceful lines and clever mechanics, but European Zonophones achieved even greater (rather quirky) elegance. Zonophones had developed a rectangular appearance on both sides of the Atlantic, a configuration especially evident in the 1904 European models, one of which is shown here. It's a conventional Zonophone, with characteristic off-set turntable, modified to accept coins, and attached to a coin box base. The customary Zonophone "S" crank, and "V Concert" soundbox are present. The horn is typical of European Zonophone models, nickel-plated brass, with red-painted interior. *Courtesy Musée de l'Aventure du Son, St. Fargeau (Value code: VR)*

3-60. Though it resembles the American model known as the "Grand Opera," this confection from the International Zonophone Company ("Luxus-Concert-Zonophone") fairly drips with sugary details. *Courtesy Musée de l'Aventure du Son, St. Fargeau (Value code: VR)*.

3-61. A similar fashionable Zonophone having inlaid wood instead of decorative nickel-plated metal appliqué. This model ("Grand-Opera-Luxus," "No. 110") sold for 250FF in the 1904 French catalog, a considerable sum. *Courtesy of Jalal and Charlotte Aro (Value code: VR)*.

3-62. In 1904 and 1905, Victor's talking machines were transitioning. One change was from letter to number model designations. Another change was from the "straight" (or front-mounted) to the "Taper Arm" (or back-mounted) horn/arm configuration. Victor's "Z" was the last of the older, straight horn models to be introduced, clearly intended to replace the low-priced "R" ("Royal"). When Victor refitted its roster with Taper Arms and numeric identification, the "R" simply disappeared, and the "Z" became the Victor "1" (also expressed as Roman numeral "I"). "Z" No. 3395, documented this transformation, since it carried a tag reading, "Remove this plate to attach Taper Arm." *Courtesy of Stan Stanford (Value code: G)*

3-63. Someone went to a great deal of trouble to create this "faux Edison." The grain-painted cabinet, the hand-done "Edison Standard Phonograph" decoration, the overall size and shape all cleverly suggest a Model "A" "Standard." *(Value code: VR)*

3-64. Within — nothing more than a well-dressed Columbia Type "B" Graphophone! Under the machine is a space, lined with the same soft cloth used inside the lid, for storing cylinder records.

3-65. The October 15, 1905 issue of the French trade journal *Phono-Ciné Gazette* ran a story entitled "Maison Dutreih" ("House of Dutreih," pronounced "doo-tray"). "...Seeing that it was founded in 1835, it is the oldest mechanical music house which produces French instruments in France. Around the time that the phonograph appeared Monsieur George Dutreih resolutely launched himself on a new path and created a brand 'Musica' which has no superior." A facsimile of a document signed by some of the most notable French recording artists of the day stated, "'Dutreih' PHONOGRAPHS are a marvel. We don't know of any other one more perfect... We have been pleased to verify, and do not hesitate to say, that to listen to a 'Dutreih' cylinder is a truly artistic diversion." It's interesting to see these luminaries endorsing with such high-flown language a product that was, quite frankly, low-end. Dutreih cylinder machines were small, with catchpenny mechanics. The best of the lot was an imitation Columbia "Eagle." Dutreih cylinders were made by Pathé. Dutreih's "Musica" disc machine line was also modestly conceived (records made by Pathé, again). Shown is the "Prima," at 31 FF. Other disc models in the 1906 catalog went up in price to 182 FF, but like the cylinder line, the cheapest were the most successful. The firm of C. Sims & Mayer of Paris employed the "Prima" to promote its *bijoux goldfilled* (gold-filled jewelry), trumpeting in an advertisement, "Free! Absolutely free! A MARVELOUS PREMIUM. We have already given away a large quantity of these phonographs and more than 50,000 other delightful premiums..." *(Value code: G)*

3-66. An engineer would find the talking machine, in its many incarnations, full of design flaws. Certainly, many points of precision were compromised for the sake of economics or simplicity. *Theoretically*, the stylus of a cylinder phonograph should maintain a straight line of contact with the surface of the record as the reproducer carriage is tracked laterally by a feed-screw. Yet, in Europe, mechanical tracking was frequently abandoned or modified in favor of "floating horn" attachments, wherein the stylus was carried forward by the record groove. These contrivances allowed free lateral movement of the reproducer (hence no skipping or running between grooves such as occurred with early Edisons or Columbias), yet the stylus traveled the surface of the record in an arc (imagine the way a Puck's reproducer meets the record), which is not the ideal. In the little instrument seen here, a clever solution to "floating horn" problem was offered. Nothing is known about it, except that it bears a certain resemblance to French machines of the 1905 period (the cabinet, Inter-sized mandrel), though the works are more likely German (note the "lyre" design of the base, suggesting a Puck). *Courtesy of Collection Fabrice Catinot, Dijon (Value code: VR)*

3-67. At the base of the motor housing can be seen a protruding metal "finger." Below this, notice a curved plate attached to the wooden plank. Rough edges may be seen on some of the components — this may have been an ingenuous machine, but it was not an expensive one. This phonograph's secret is explained by the next illustration. *Courtesy of Collection Fabrice Catinot, Dijon*

3-68. As the machine plays, the "finger" periodically engages a "claw," which drags the entire motor housing in an arc along the teeth below. Consequently, the mid-point of the cylinder record changes location as it plays, and the reproducer's stylus travels the record in a straight line. *Courtesy of Collection Fabrice Catinot, Dijon*

3-69. During the period 1905-1909, this Gramophone enjoyed great popularity in Europe. Known as "Monarque No. 9" in the French catalog (remember, the model numbers were reused), it also appeared under the distributorship of other European Gramophone agencies. Here it is seen in its earliest version, with brass motor, record securing nut and nickel-plated straight horn; later examples incorporated the improved worm gear motor and a *pavillon volubilis* (literally "morning glory" horn, more commonly known as a "flower" horn in the United States). *Courtesy of Bert Gowans (Value code: G)*

INSTRUCTIONS.

To change the tune push bar _gently_ to the left until number on dial corresponds with number on programme you wish to hear, _then_ drop a nickel in the slot, but not while the machine is running.

1 2 3 4 5

PRO

3-70. When coin-operated talking machines first appeared, their novelty was sufficient to overcome the drawback of one record selection per machine. Exhibitors changed records frequently, as attested to by marquee signs from nickel-in-the-slot instruments which were hurriedly hand-produced to bring timely selections to the patrons. Inventors, however, dreamed of creating _multiple_ choices for the discriminating listener. One of the rarest of the discrete systems was the "Multiplex Coin-in-Slot Machine." In 1905, Great Britain's Edison Bell company listed this apparatus for £10 10 0. Edison Bell had a rather ambivalent relationship with Edison's National Phonograph Company in the United States. Edison Bell had the exclusive rights to exploit Edison's Phonographs in Great Britain until the middle of 1903, after which its connection to the American "parent" firm became muddy. Thomas Edison devoutly wished that Edison Bell would just go away, so he could take over the British market without competing against his own name. Edison Bell, however, was not about to be dislodged — in fact it flouted Edison's hegemony, "The Phonograph has been introduced so frequently that it has become familiar to everyone… Everyone knows Edison invented it… Everyone **does not know,** but every Britisher **ought to know** — that THE PHONOGRAPH IS NOW A BRITISH PRODUCT… **'Wake up, England!'** — Prince of Wales." The "Multiplex" contained five two-minute cylinder records on a carousel. The large lever at the front of the cabinet allowed the patron to make his choice. The "Multiplex attachment" to Edison cylinder Phonographs had been around since 1896, when it was patented by George W. Moore. Moore had a short-lived partnership with George V. Gress, but the device remained at the periphery of the talking machine market until taken up by Edison Bell. (see _Discovering Antique Phonographs_, Fabrizio & Paul, figure 2-20.) _Courtesy of Sam Sheena (Value code: VR)_

3-71. The first conception of an "internal-horn" talking machine was rather literal. The German "Hymnophon" of 1905 encased an unaltered horn in its cabinet. The bell protruded conspicuously. This may seem like a primitive design, haphazardly created during the first, faltering steps of the genre — however, it was also a matter of taste. Five years later, the Pathé "Jeunesse" embraced a similar embedded-bell configuration. The November 15, 1905 issue of the French trade publication *Phono-Ciné Gazette* featured a "Hymnophon" advertisement, from Ernest Holzweissig's Successors, Leipzig. The Victor/Gramophone trade mark was slyly suggested by a group of dancing Dachshunds, one of which was on his haunches, with ears pricked-up, in front the instrument's jutting bell. A French patent was shown, No. 344,328, and the Paris distributor was listed as *Dépôt de l'Hymnophon-Pantophone*. In fact, the "Hymnophon" also was sold in France during 1905-1906 under the name "Pantophone." The language that would lure the public away from external-horns was already in place in the "Pantophone" catalog, "Without visible horn…An absolutely solid idea, the 'Pantophone' is the most easily manageable of all existing devices. With it, no horn to remove and pack up separately, no cumbersome carrying cases for the horn, machine and accessories; the whole thing can be closed up in one case of small dimensions and easy portability." *Courtesy of Wilfried Sator (Value code: VR)*

3-72. An unpretentious suggestion of the "Hymnophon," but in the cylinder record format, was this French "Stylophone." Not much to look at from the outside. *Courtesy of the Julien Anton collection (Value code: VR)*

3-73. Within, an ordinary Graphophone "Eagle" with an "internal" horn that snaked up above it. Herzog's "Cylo-phone" of 1908 was a giant-sized version of this same concept (see figure 3-137). *Courtesy of the Julien Anton collection*

3-74. To hide the purpose of an object might be motivated by the desire to conceal something rude — such as fancifully embellishing a coal scuttle, or secreting it inside an artistically designed cupboard. Yet, disguising household articles can be merely a game inspiring respect for the cleverness of the sham. This French cylinder record storage box has been made to look like a row of books. It may sit on a shelf, with its practical use unknown. If one were to look closely, however, the spines are marked "Phonograph Cylinders" and "Patented" (in French). Note the Phrynis records (see figure 3-151). *Courtesy of Collection Fabrice Catinot, Dijon (Value code: J).*

3-75. Edison promotional posters of the 1903-1906 period were generally colorful, and often pictured women and children enjoying recorded music. This example fits the general mold, but interestingly depicts a Japanese girl in whose tiny hands the cylinder record takes on a rather large appearance. The artist seems to have struggled with proportion, as the featured Edison "Home" Model "A" looks incapable of accommodating the girl's mammoth record! Measures 18 1/2" x 26" unframed. *Courtesy of Brice Paris (Value code: VR)*

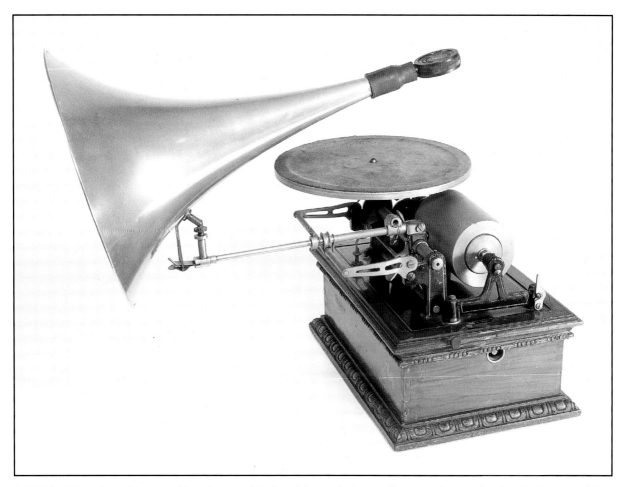

3-76. The idea of a talking machine that would play either cylinder or disc records remained active in the public imagination throughout the first decade of the twentieth century. There were commercially available cylinder/disc contraptions such as the Idéal (see *Discovering Antique Phonographs*, Fabrizio & Paul, figure 3-31), as well as attachments for converting ordinary instruments to double duty (see *Antique Phonograph Gadgets, Gizmos and Gimmicks*, Fabrizio & Paul, figure 2-46). One-of-a-kind "science projects" have been discovered, too. Where does this Pathé "No. 2" cylinder phonograph modified to include a disc turntable (suitable for playing vertical cut records) fit in? *Courtesy Musée de l'Aventure du Son, St. Fargeau (Value code: VR)*

3-77. The adaptation, which takes power off the customary Pathé drive train, is so expertly done that one suspects it isn't *bricolage* (handyman work). It would have been accomplished in 1906, when Pathé introduced its first disc records, which were soft wax and played by a conventional Pathé cylinder-style reproducer. We fear the word "prototype," since it has been used to excuse a sideshow gallery of monstrosities — but in this instance we do not demur. *Courtesy Musée de l'Aventure du Son, St. Fargeau*

3-78. The Lorelei (originally Loreley) is a large rock that juts out of the river Rhine, on the eastern bank near St. Goarshausen, Germany. A legend came to be associated with it, of the spirit of a maiden who beckoned sailors too near the shallows with her singing. The "Lorelei" Puck, shown here, suggested the rocks and the waves, and the water wraith with a harp to accompany her enticing voice. Talking machine dealer Vittorio Bonomi of Milano, Italy, advertised the "Fonografo 'Loreley'" for 20 Lire, circa 1906 — "The device is supplied with an acoustic arm, and swiveling lily horn of 26 centimeters, elegantly enameled." The overall height of this example is 19". *Courtesy of Jalal and Charlotte Aro (Value code: VR)*

3-79. Female figures graced several "Puck" phonographs, though perhaps not in the most complimentary context, since all were "sirens," including two types of mermaid and one Lorelei. Except for this rarity, that is. In this instance, she is a less dangerous musician. *Courtesy of the Julien Anton collection (Value code: VR)*

3-80. Inspired by the late Symbolist art of the period, reminiscent of the work of Gustav Klimt, and especially of Jan Toorop, the lady harp player transcends the humble industrial form of a Puck. *Courtesy of the Julien Anton collection*

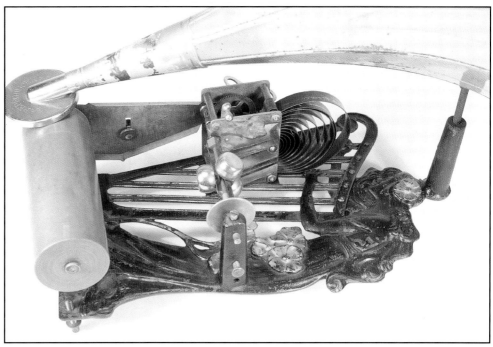

3-81. Puck cylinder phonographs were generally defined by their simplicity, and similarity to clock mechanisms. Beyond that, there was a great deal of variety in aesthetic design. In the category of machines known as "Kastenpucks" (Pucks with wooden boxes), the greatest diversity is found. Seen here, an instrument capable of playing an "Inter" cylinder, featuring a bas relief of an old Teutonic hunting seen, glazed in subtle colors. *(Value code: G)*

3-82. The Rhineland hunting scene is washed with mellow, translucent hues.

3-83. In 1906, Pathé, a huge success in the cylinder talking machine business, introduced disc instruments. The company's first models employed a system of vertical cut reproduction that was based on Pathé cylinder technology. Note, in this early Modèle "A," the horn and ebonite reproducer are practically identical to those already in use on Pathé cylinder machines. The record itself comprised a heavy composition backing (akin to cement) on which vivid company advertising was printed (shown), and on the reverse of which was affixed a recording pressed in rather fragile wax. The ephemeral recorded format was soon discovered to be vastly inferior to conventional shellac from which Pathé discs were soon being pressed. The "dust cover" for the turntable was a fine idea, but rarely exploited by talking machine firms. Note the rooster logo, carried over from the cylinder line.
(Value code: VR)

3-84. Inside the dust cover was a set of instructions. Although disc talking machines were fairly common by 1906, the Pathé system was sufficiently specialized to suggest the need for guidance.

3-85. Pathéphones during the 1906-09 period were identified by letter designations, after which they were numbered. This Modèle "S" was at the top of the early line, with sleek Sheraton styling, and an immense blue horn. Although the *very* first Pathéphones carried over the rooster logo from the firm's cylinder instruments, a decal featuring a discus thrower preparing to toss a Pathé disc was later introduced. It can be seen here atop the wooden dustcover. *Courtesy of Jalal and Charlotte Aro (Value code: VR)*

3-86. The formidable motor of the Pathéphone "S." Pathé motors were well-engineered and powerful, as good as or better than Victor or any other firm with a reputation for quality. *Courtesy of Jalal and Charlotte Aro*

3-87. The Douglas Phonograph Company, located in New York City, made a specialty of supplying high quality auxiliary cabinets to owners of talking machines, particularly Victors. The owner of this Victor "VI" ($100.00) availed himself of one of the Douglas "Marquetrie" cabinets. Two shelves for record storage were hidden behind the doors. With the appearance of the Victrola in 1906, Douglas's business in "talking machine furniture" began to wane. Cabinet measures 21" wide, 16" deep, 47" high. *Courtesy of Brice Paris (Value code: VR)*

3-88. In 1907, the internal-horn talking machine was brand new. In the United States, Victor's "Victrola" was just beginning an inexorable rise to undreamed-of prominence, and in France Compagnie Française du Gramophone, one of Victor's European affiliates, was offering the "Amphion." The name "Amphion," meaning internal-horn Gramophone, was used because "Victrola" would have made no sense in Europe. Period furniture styles were offered, including this one, much influenced by Sheraton. *Courtesy of Phonogalerie, Paris (Value code: E)*

3-89. For those familiar with the "conventional" form of a Victrola, the logical place for the lid to open is below the sweep of the upper cabinet, at the place where the sides become straight. However, before this wisdom had been established, both the American "Pooley" Victrola and the European "Amphion" opened rather astonishingly above. *Courtesy of Phonogalerie, Paris*

3-90. In 1907, court decisions forced Edison to remove certain features from the designs of his Phonographs, and the modified instruments were designated Model "C." These reconfigured Phonographs were available first in New York State (as mandated by the court), and slightly later in other markets, such as Philadelphia, where this "Gem," sold by James Bellak's Sons, was emblazoned ostentatiously. *Courtesy of Stan Stanford (Value code, ordinary Model C Gem: I)*

3-91. The oddly shaped "Tulip" horn was protected by the United States Design Patent No. 38,447 (Feb. 19, 1907) of Morris Geller, a Russian citizen living in Newark, New Jersey. It was offered by the New Jersey Horn Company in 1907. This version for a disc talking machine has a 21" diameter bell, and displays a spray of white mums on a field of powder blue. *Courtesy of the Johnson Victrola Museum (Value code: VR)*

3-92. Colorful decals would surely have enlivened the look of sedate American Victor talking machines. Yet, Eldridge Johnson chose the conservative road — no ostentatious decals! G&T in Great Britain, however, saw nothing wrong with embellishing the front of this otherwise modest instrument (circa 1907, known as the "Baby Monarch," essentially an export Victor "I") with a cartoonish banner featuring the "His Master's Voice" logo beneath "The Gramophone." *Courtesy of Philippe Le Ray (Value code: G)*

3-93. And while on the subject of dressed-up Victor "I"s, this charming little instrument from the Deutsche Grammophon catalog of the same period incorporated conventional Victor components into a pressed wood cabinet. The horn is the color of "grass green" that had enduring popularity on Gramophones in Europe and especially Great Britain. The round celluloid tag on the front corner of the cabinet displayed the "Recording Angel" logo. *Courtesy of Richard and Jill Pope (Value code: G)*

3-94. The Talkophone Company, like its competitors, offered a line of machines to fit most any budget. The various models were named after prominent band leaders of the day, commencing with "Herbert" (as in Victor Herbert) and rising in cost, features, and status. The top of the Talkophone line (initially $75.00; later reduced to $40.00) was, naturally, the "Sousa" — John Philip was, after all, the "March King." Late in the life of the firm (1907-08), it produced some unmarked models (as shown here) which suggested the elegant appearance of the "Sousa," but were actually created with cheap appliqués. *Courtesy of the Scott and Denise Corbett collection (Value code: G)*

3-95. This patent notice was pasted inside an example of the earliest version of the Victor Victrola (as identified on the ID plate: "VTLA," serial number 1110). Considering that "Auxetophone" referred to an entirely different type of apparatus (an external-horn, compressed air activated loud speaking device), it's highly unusual to find the term inside an otherwise standard-issue "VTLA." A mere mistake by the label paster? A last-minute substitution when the "VTLA" labels temporarily ran out? Caution is the best companion when analyzing an anomaly — furthermore, at least one other "VTLA" in the same serial range has been discovered with this notice affixed. The true reason for the contradictory labeling may be lost in time. *Courtesy of Bob Thomsen*

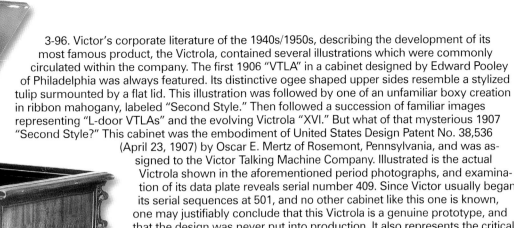

3-96. Victor's corporate literature of the 1940s/1950s, describing the development of its most famous product, the Victrola, contained several illustrations which were commonly circulated within the company. The first 1906 "VTLA" in a cabinet designed by Edward Pooley of Philadelphia was always featured. Its distinctive ogee shaped upper sides resemble a stylized tulip surmounted by a flat lid. This illustration was followed by one of an unfamiliar boxy creation in ribbon mahogany, labeled "Second Style." Then followed a succession of familiar images representing "L-door VTLAs" and the evolving Victrola "XVI." But what of that mysterious 1907 "Second Style?" This cabinet was the embodiment of United States Design Patent No. 38,536 (April 23, 1907) by Oscar E. Mertz of Rosemont, Pennsylvania, and was assigned to the Victor Talking Machine Company. Illustrated is the actual Victrola shown in the aforementioned period photographs, and examination of its data plate reveals serial number 409. Since Victor usually began its serial sequences at 501, and no other cabinet like this one is known, one may justifiably conclude that this Victrola is a genuine prototype, and that the design was never put into production. It also represents the critical stage where Victor designers were considering the elimination of the curvaceous elements that had appeared in the Pooley design. Evidently, the rather stark, boxy appearance of the Mertz cabinet here illustrated didn't please the brass at Camden. At that point, the decision was made to retain Pooley's original ogee upper sides, but to *invert* them and incorporate the sides into the lid, thus forming a dome. In this way, the iconic Victrola lid was born. *(Value code: VR)*

3-97. This otherwise unremarkable Victor "V" features a Hawthorne & Sheble flower horn of 1907 design (United States Design Patent No. 38,478; granted to Horace Sheble, March 12, 1907). Cabbage rose decals are featured as well as the firm's "Crystal" effect of translucent paint over galvanized steel. *Courtesy of Mark Hansen (Value code, ordinary Victor V with flower horn: F)*

3-98. A close-up of Hawthorne & Sheble's "Crystal" finish, subtly forming the background for embossed panels, pinstripes, and floral decoration. *Courtesy of Mark Hansen*

3-99. The Type "C" Graphophone appeared in 1897 as the "Universal," capable of playing for an hour with one winding of its powerful triple-spring motor. For those preferring a battery-powered 2 volt electric motor, the Type "CE" was available, and also a variant for use with 110 volt direct current (Type "CI"). This example, the Type "CA," for use with alternating current, is the most unusual because the AC system for homes and businesses championed by Westinghouse had yet to gain widespread popularity. The serial number suggests that this was the sixth mechanism built in what must have been a very limited run prior to the introduction of metal-cased Commercial Graphophones and Dictaphones. *Courtesy of Don Courtesy of Don Fenske (Value code: F)*

3-100. The Type "CA" Graphophone's primordial Westinghouse motor for alternating current. *Courtesy of Don Courtesy of Don Fenske*

3-101. The "Klingsor" brand had an unusually long life. It began in Germany, in 1907, and ended up being promoted in Great Britain during the 1920s. "Klingsor" had a couple of distinctive features: a spruce sounding board on which a tuned harp was mounted (becoming, "the disc talking machine with 'String-Resonance'"), and stained glass doors. Both can be seen here in this highly unusual floor model. Although most "Klingsors" were table models, it's interesting to note that in the 1907 catalog issued by the manufacturer, Stephan Hain of Krefield, Germany, two floor-standing instruments were shown. This particular device probably dates from a year or two later. At that time, "Klingsors" were being assembled with Pathé motors and turntables. Experience has shown that during the early years of the marque a large percentage of "Klingsors" were made up as coin-ops. Though many have lost their coin-controlled equipment over the years, by accident or neglect, some were modified at the factory to be sold for domestic purposes. It's clear that "Klingsor" was very much a brand intended for public amusement. *Courtesy of Bill Boruff (Value code: VR)*

3-102. Excelsior was one of the most prolific of the German brands. The company worked to develop a specialized market in Great Britain. Excelsior offered to the British public a series of cylinder phonographs bearing the names of gemstones ("Ruby," "Pearl" — though a pearl is not a stone…). Here we see the better-quality "Diamond," which has a 6" long mandrel to accommodate the special 6" long cylinders of the 1906 period. (There were Columbia "Twentieth Century," Lambert "Imperial," and Excelsior's own version of these records.) This example has a back-mounted horn configuration. Excelsior instruments can often be identified by their characteristic lid handles, even if no company markings are present. *Courtesy of Jürgen Bischoff (Value code, back-mounted version: VR)*

3-103. The British branch of the Lambert Company (celluloid cylinder record manufacturers) sold phonographs, the American branch did not. Lambertphones were German made, from all appearances by Excelsior. The "Companion" was an inexpensive "reversible" model (the works inverted to store inside the cabinet). *Courtesy of Mark and D'Arcy Gaisser (Value code: H)*

3-104. The reproducer employed a special articulation called a Rawlinson Joint. Note the original winding crank which is decorated to match the machine's motif. *Courtesy of Mark and D'Arcy Gaisser*

3-105. To be certain, even the most die-hard Victor partisan eventually could grow bored of the same Victor "I, II, III, IV, V, VI" domestic line-up. Then why not feast one's eyes on the variety offered worldwide by the Victor/Gramophone affiliates? Deutsche Grammophon Style "No. 14" embodied extraordinary mahogany and brass cabinetry, a triple-spring Victor motor, and an imposing nickel-plated brass horn. The same cabinet had been available with the front-mounted "Monarch No. 11a de Luxe" advertised in October 1902. *Courtesy of Jürgen Bischoff (Value code: VR)*

3-106. Being out of the reach of Mr. Edison's tireless attorneys was a wonderful thing. Many American firms (such as the U-S Phonograph Company of Cleveland) would have wished for no less. However, it was in Europe that the long arms of Tom's "Phonograph Chasers" ran out of length. Therefore, a German firm could produce an Edison "Standard" knock-off (seen here), and even insolently label it "Edison." A similar instrument was offered by Alfred Kirschner & Co. of Berlin, in 1908. *Courtesy of Jürgen Bischoff (Value code: H)*

3-107. The reproducer of the imitation Edison. It appears to be loosely based on the Edison Model "B," with an added weight appended, reminiscent of the way that Edwin Mobley adapted ordinary Edison reproducers, coaxing increased volume out of them by affixing a protruding weight. *Courtesy of Jürgen Bischoff*

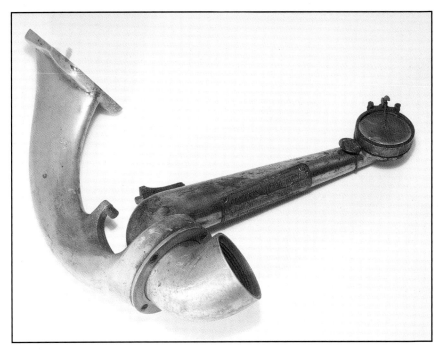

3-108. The American Graphophone Company's factory in Bridgeport, Connecticut, contained an archive which documented the firm's innovations and supported the company's frequently advertised assertion that it had created the talking machine business. Some of the exhibits had been used during endless rounds of litigation; others were prototypes of defining Columbia Graphophone products. Most items in the archive were marked by a rectangular plate with a reference or "exhibit" number. Columbia was heir to this precious collection of artifacts, which moldered for decades after its value was no longer appreciated, and was finally scattered to the four winds. We see here a remnant from the Columbia archive — a prototype of the back-bracket, threaded elbow, cast tone arm and "Analyzing" soundbox used on thousands and thousands of Columbia Disc Graphophones of the 1905-1910 period. Perhaps the unusual fitting at the base of the arm (to lessen drag on the record) was being considered. *Courtesy of Mark and D'Arcy Gaisser (Value code: VR)*

3-109. Models of components under development were usually made of bronze. Here we see a polygonal tone arm, never put into production, a prototype of an arm customarily used on later (1914) external-horn Disc Graphophones, and a later style threaded elbow to which a peculiar, curved spout has been affixed. *Courtesy of Mark and D'Arcy Gaisser (Value code: VR)*

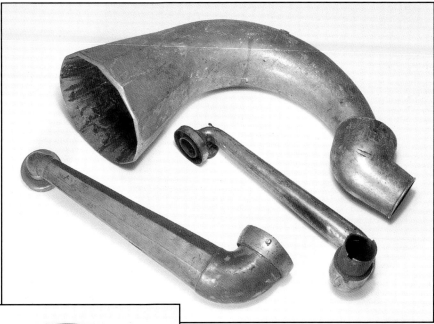

3-110. At one time, Columbia was in possession of the molds from which its vast catalog of two-minute cylinder records had been produced. Most of these industrial artifacts passed into oblivion. The group shown here is the largest number we have seen in one place. *Courtesy of Mark and D'Arcy Gaisser (Value code: VR)*

3-111. The Columbia Graphophone Type "BF" ("Peerless") was introduced in September 1905. This pristine example has more to commend it than its fine state of preservation. Note the absence of a winding crank, or any place to insert one (commonly on the right side). *Courtesy of Jerry Blais (Value code: VR)*

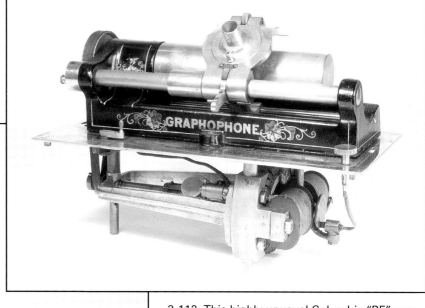

3-112. This highly unusual Columbia "BF" was powered by a 2 volt battery motor. Certain Graphophones had been available with similar low-voltage motors since 1893. It's extraordinary to consider that this type of motor, which drove the very first Graphophones after the foot treadle era, was employed in this "modern" instrument, with 2 and 4 minute gearing (note attachment protruding at left). *Courtesy of Jerry Blais*

3-113. An advertising mirror for "Graham & Wells, Druggists, Corvallis, Ore., … Agents for Edison Phonographs… Kodaks a Specialty." This promotional piece was made by Stanford-Crowell Co., Ithaca, N.Y., Ithaca Sign Works, Advertising Signs and Novelties. *Courtesy of Jerry Blais (Value code: J)*

3-114. The 1908 Columbia "Special Spring-Tension Reproducer" was equipped with a stylus designed to play either two-minute or four-minute "Indestructible" cylinder records. This was accomplished through the use of a steeply tapered stylus, effectively sized slightly larger than a typical four-minute, but slightly smaller than a typical two-minute stylus; splitting the difference. Not recommended for rare or valuable cylinders! *Courtesy of the Scott and Denise Corbett collection (Value code: J)*

Among collectors of coin-operated talking machines, the Multiphone and its multiple-cylinder playing descendants hold particular appeal.

A phonograph mechanism employing a vertically-mounted wheel or carousel to hold multiple cylinder records was originally patented in the United States by Cyrus C. Shigley on May 5, 1903 (No. 727,002). Shigley subsequently was granted four other patents relating to selective, coin-operated cylinder phonograph apparatus. His No. 841,727 (granted January 22, 1907) was assigned to the Multi Phonograph Company, Incorporated of Grand Rapids, Michigan.

This firm manufactured the first versions of the Multiphone, which by 1906 had evolved into a behemoth employing 24 two-minute cylinders. A separate entity, the Multiphone Operating Company of New York City, offered stock to investors in November 1906. By May 1908, this firm was in receivership, and was subsequently reorganized.

Shigley left the company and began another firm in Kalamazoo, manufacturing the Kalamazoo Electric Phonograph, a similar 24-cylinder-playing mechanism. On a separate front, a New York City businessman named Julius Roever had filed, on May 29, 1907, for two patents relating to "Multiple [record playing] Phonographs" (granted as 883,970 and 883, 971, both April 7, 1908). Examining the Shigley and Roever patents reveals a common approach: both men championed the vertical, cylinder-carrying carousel. This is was no coincidence — Mr. Roever was the

general manager of the Multiphone Operating Company, which was responsible for the day-to-day exhibiting of the machines, east of the Mississippi and north of the Mason-Dixon line.

A third man patented a similar design at about the same time — Cornelius Reinhardt of San Francisco was granted No. 909,455 on January 12, 1909. This patent was assigned to the Autophone Company of San Francisco. It would in fact be an "Autophone" that would become the descendant of the Multiphone, but from a firm called the American Phonograph Company, located in Brooklyn, and run by Julius Roever.

Around 1910, Roever assumed control of the struggling Multiphone enterprise. Under the auspices of Julius Roever, new products were developed that encompassed both the sublime and the ridiculous. Roever's patents were employed to create various versions of a multiple-cylinder playing device for home use called the Autophone. However, as the firm's prospects waned, the Autophone name was also applied to a line of conventional disc talking machines, steadily diminishing in quality. These bore little resemblance to the monolithic mechanisms that spawned them. Through the fortuitous discovery of a cache of Multiphone/Autophone documents, this progression can now be illustrated for the first time. (Note: no values are applied to these one-of-a-kind documents.)

3-115. The giant of the Multiphone "family": an eight-foot leviathan containing 24 cylinders arranged on a revolving "Ferris wheel" magazine behind plate glass. The "Bronze Cabinet [finish]" illustrated has not been reported. (See *Discovering Antique Phonographs* by Fabrizio & Paul, figure 3-88 for a photograph of a Multiphone in a similar cabinet of mahogany.) The mechanism was driven by a three-spring motor similar to the "Triton" that ran the Edison "Triumph" Phonograph.

The Multiphone

1907

Bronze Cabinet

8 FEET HIGH, 3 FEET 5 INCHES WIDE
18 INCHES DEEP

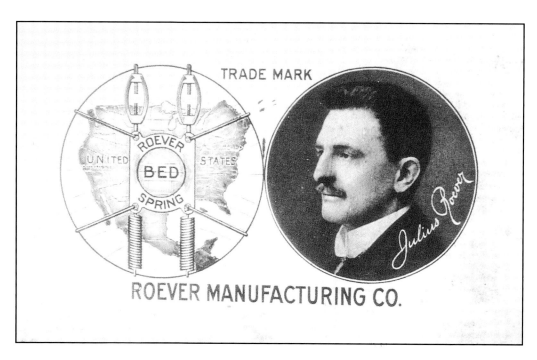

3-116. Julius Roever, whose inventiveness encompassed such diverse items as bedsprings and talking machines.

3-117. The Multiphone "family" gathered outside its plant.

3-118. This interior shot of the Multiphone plant shows various components of the machines, most notably the record magazine wheels in the foreground.

3-119. On September 26, 1907, employees and wives gathered at the Multiphone Company picnic. A band was in attendance, as well as one of the company's machines.

3-120. This photograph from a long-ago summer day commemorated a company river cruise on the side-wheeler *Isabel.*

3-121. By 1913, Roever had formed the American Phonograph Company, which was manufacturing a scaled-down version of the Multiphone called the Autophone. This Type "A" could accommodate 12 cylinders in its circular magazine.

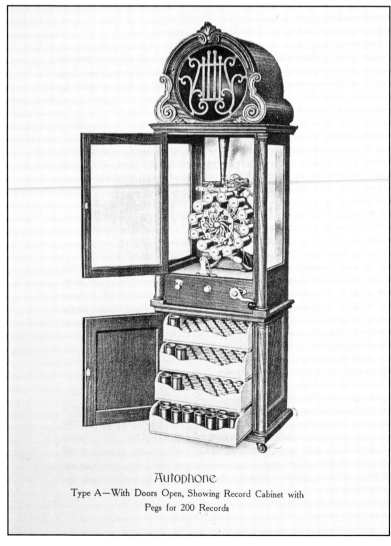

Autophone
Type A—With Doors Open, Showing Record Cabinet with
Pegs for 200 Records

3-122. This view of the Auto-phone plant interior shows employees assembling the machines. It is not clear if this was done in the Brooklyn plant or at 102 West 101st Street in Manhattan. The latter location may have been limited to offices and showrooms.

3-123. The best-known version of the Autophone for home use. Some examples exist and show the machine to have been a high quality product, though the use of innovative but unstable die-castings has caused some modern difficulties. See *Discovering Antique Phonographs,* by Fabrizio & Paul, figure 4-23 for a photograph of the Autophone.

3-124. Until the discovery of this sample cylinder record label, it was not known that Roever had considered developing a private brand for records. It is unlikely that this line of records was ever produced. The timing was particularly bad, since the term "Diamond Record" in 1913 would have run afoul of Edison's soon-to-be introduced "Diamond Amberolas" (Models "30," "50" and "75"), as well as his emerging Disc Phonograph terminology. (There also had been "Diamond Records" [lateral cut discs] marketed in Chicago a few years earlier.)

AVTOPHONE
DIAMOND
RECORD

Autophone Mfg. Corp. N.Y.

THE AUTOPHONE
PLAYS 12 RECORDS
AUTOMATICALLY

3-125. This interior view of the Autophone plant would be unremarkable except for the cabinets lined up against the far wall.

3-126. Looking distinctly cylinder Autophone-like is a *disc* version of the brand. Note the characteristic soundbox designed for vertical or lateral cut records (it is set in the vertical cut position). Breaking from tradition, the Autophone's record storage was placed above the speaker grille, a configuration employed by another iconoclast, "Puritan."

3-127. This disc Autophone sports a more elaborate cabinet in what appears to be an antiqued finish.

3-128. Three high-end disc Autophones which illustrate the fine cabinetry still offered by Roever's firm. The example on the right is evocative of the Edison Disc Phonograph Model "A-450."

3-129. Most talking machine companies, when developing a line of table model instruments, offered more expensive designs with lids, and less expensive models without that feature or other frills. The Autophone dispensed with any pretense of a high quality table model, marketing only four simple designs in inexpensive oak finish (actually chestnut) or mahogany finish (actually birch) cabinets. This was the top of Autophone's table model line. The soundbox is adjusted for lateral discs.

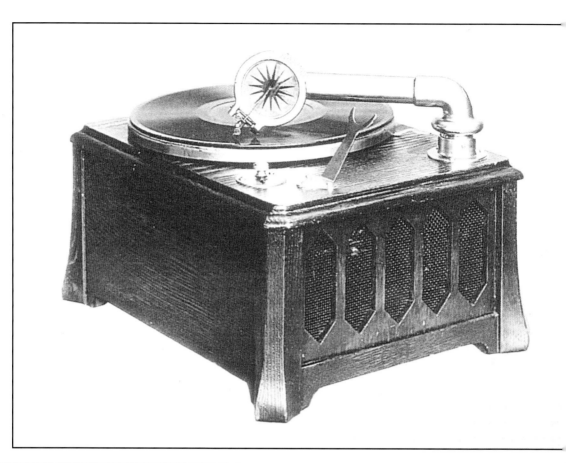

3-130. The end of the Autophone line. These three cheap table models appear to have been the forlorn finish of Julius Roever's talking machine enterprise. In a world awash with the $15.00 Victrola "IV," the elaborate Autophones of earlier days could not compete. A single advertisement for the Autophone appeared in the February 1919 issue of the *Talking Machine World.* It may be presumed that the Autophone table models, descendant of the elaborate disc and cylinder models, and grandson of the mighty Multiphone, disappeared around 1920. (See *Discovering Antique Phonographs*, by Fabrizio & Paul, figure 4-82, for a photograph of a surviving table model Autophone.)

3-131. A most unusual product of the Multiphone/Autophone Company was this coin-operated machine, incorporating a wheel of 12 cylinders within a glass dome. A brass plate reads, "This Machine Is The Property of Wolverine Furniture Co., Zeeland Mich[igan], U.S.A." *(Value code: VR)*

3-133. A view with the glass dome removed shows details of the mechanism. Note the feedscrew to the left and the coarse return screw above it. As the reproducer is driven toward the front of the machine during play, the bent tube above it telescopes outward while directing the sound into the nickel-plated horn to the rear.

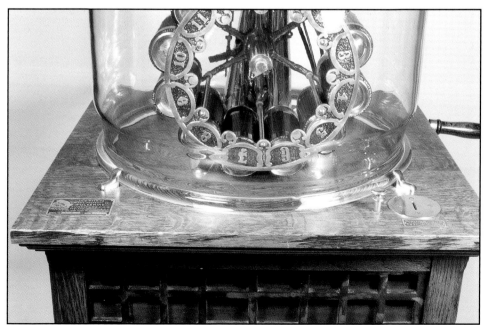

3-132. The coin-slot may be seen at the right, and the brass identification plate to the left. This device was manufactured under United States Patent No. 909455 (granted January 12, 1909), and is serial No. 106, Model "SB." The patent was the applied for on December 6, 1907 by Cornelius Reinhardt of San Francisco, and when granted, was assigned to the Autophone Company of San Francisco, a California Corporation. Considering that Roever was located in Brooklyn, it's interesting that the Reinhardt patent was assigned to an "Autophone" firm on the opposite side of the country.

enormous3-135. Anyone who has studied contemporary photographs of early twentieth century ladies might conclude that they frequently had big aspidistras. Ferns, and other houseplants were popular, too. So, why not a talking machine disguised as foliage? The "Floraphon," an imaginative German production, was innocuous when not in use. 43" overall height. *Courtesy of Wilfried Sator (Value code: VR)*

3-136. The Type "B" Graphophone, or "Eagle," is considered a relatively early (introduced 1897), exclusively two-minute machine, the last of which were built in 1906. However, this example is an exception. It was an early production model (as attested by the plates linking the spring barrels), updated a decade after it was manufactured in order to play four-minute cylinder records (which appeared in 1908). The professionally made gear train mounted to the front of the motor plate allowed the unusually small stylus of the modified reproducer to play the rather incongruous Edison four-minute "Blue Amberol" seen on the mandrel. It is a mystery why someone took such trouble to modify an inexpensive ($12.00 when new), aging mechanism. *Courtesy of the collection of Howard Hazelcorn (Value code, this particular modified machine: VR)*

3-134. This page from the Spiegel general catalog of 1908 documents an interesting and previously unknown variant of the Columbia "client" machine that was usually sold as a Standard Talking Machine Style "X." The commonly seen configuration of this instrument is shown at the bottom left, here dubbed "The Excelsior." Above, is shown the same machine with a conventional Columbia back-mounted horn arrangement. Yet, note the positioning of the components which puts the winding crank at the front, a place it occupied only in certain German talking machines, such as the "Klingsor." Was the "Superior" really sold in this orientation (we have seen none), or was this merely a catalog artist's conception? *Courtesy of Harvey P. Kravitz*

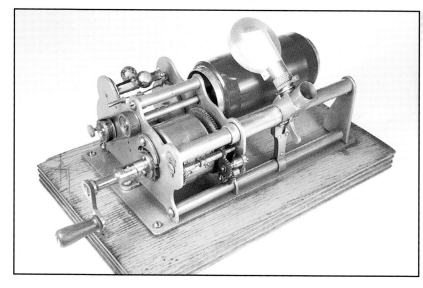

3-137. The "Cylo-Phone" (pronounced *sillo,* as in *cyl*inder) was a spectacular product of the Herzog Art Furniture Company of Saginaw, Michigan. Herzog created some of the most unusual (and stoutly-made) record cabinets of the twentieth century's first decade, but this particular version of the "Cylo-Phone" exceeded all others in elegance. The delicate, floral details are juxtaposed against robust, quartered oak — making the piece at once both graceful and grounded. The owner sequestered his non-descript cylinder phonograph within, and hooked it up to a horn that directed the sound out through the decorative grille. It was introduced in April 1908, but lasted only about a year in production — making it all the more extraordinary today. *(Value code: C)*

3-138. The Hawthorne & Sheble (pronounced "Sheb-lee") Manufacturing Company was a prominent producer of talking machine horns, record storage cabinets, and other accessories from 1895 until 1909. In 1907, with the demand for aftermarket horns diminishing, Hawthorne & Sheble began to manufacture and sell a complete line of high-quality "Star" disc talking machines equipped with clever devices, some of which were calculated to circumvent the Victor patents that would eventually choke the life out of the company. In audacious fashion, the firm offered a line of *ten* external-horn "Star" machines (priced from $10.00 to $75.00), and in December 1908 added two Victrola-inspired internal-horn models, which H&S christened the "Starola" (priced at $175.00 and $250.00). As noted, the "Star" line was intelligently engineered and the more expensive machines exuded quality. The machines included features found on no other talking machine, such as a quickly detachable tone arm, spring and gravity-driven "Yielding Pressure Feed" (which purportedly used the record groove for *restraining* rather than *propelling* the needle), an adjustable soundbox with interchangeable diaphragms, a combination soundbox support and spent needle receptacle, and a "tone modifier" located within the tone arm. Unfortunately, all these advantages coupled with superior cabinetry, over-built spring motors, and large grain-painted horns of heavy gauge steel came at a price. "Star" talking machines were more expensive than comparable Victors. Furthermore, Victor was not deterred by sly H&S technology — the lawyers from Camden descended. In rough straits, but still plucky, H&S held out during incessant litigation. However, by July 1909, Hawthorne & Sheble was bankrupt, and the innovative "Star" talking machines, after less than two years on the market, were no more. *(Value code: D)*

3-139. This unusual "Star" machine does not appear in any known catalog or advertisement of the 1907-1909 period. The instrument sports a 28" long embossed metal horn (patented by H&S in 1907), grain-painted to match the cabinet, and is powered by an unconventional motor whose two spring barrels are separated by 3", flanking a centrally-mounted governor. Unlike other known "Star" machines, this example does not carry a star-shaped data plate, merely a rectangular, nickel-plated tag stamped with a list of H&S patent dates. It is possible that this machine was one of a small group manufactured especially for a retailer to sell under a different name (H&S is known to have produced "client" instruments; see *Discovering Antique Phonographs*, Fabrizio & Paul, figure 3-103). More likely, it was among the last machines assembled, and was sold by H&S without attribution and — significantly — without serial number. Victor had secured a preliminary injunction on June 4, 1909, and Hawthorne & Sheble would have been anxious to dispose of its remaining inventory — with no serial numbers to suggest production dates — before a permanent injunction was ordered. In either event, this unique mahogany talking machine represents one of the last products created by a pioneering force in the industry.

3-140. This unique talking machine was discovered in the early 1970s, and subsequent study has revealed nothing of its origins. The machine sits on a base measuring 9 1/2" x 11". Four small legs at each corner suggest that the instrument was not meant to have a cabinet. The presence of a return screw, to reset the carriage, points to the possibility of coin operation, yet there is no automatic stop; the motor will run until its twin mainsprings are at rest. The machine is capable of playing both two and four-minute cylinders, indicating a date of late 1908 or thereafter. The device was found with an Edison Model "H" (four-minute) reproducer fitted to it. The entire mechanism was professionally designed and executed, even featuring jewel governor bearings, such as those employed in music boxes. Beyond these observations, this talking machine remains an enigma. *Courtesy of Sam Sheena (Value code: VR)*

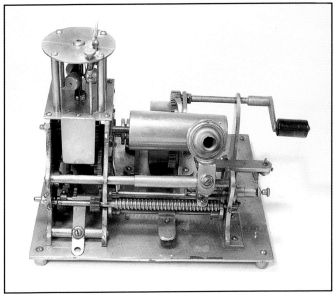

3-141. This Edison "Home" Model "A" sports a Model "D" repeating attachment as well as a Munson collapsible horn decorated with white cabbage roses. The horn measures 31" long, 21" in diameter. The unusual cylinder record cabinet employs two pivoting doors, whose interior sides each hold 40 cylinders (80 total). Cabinet measures 20" wide, 15" deep, 36" high. *Courtesy of Bob and Karen Johnson (Value code, this outfit with Munson horn: VR)*

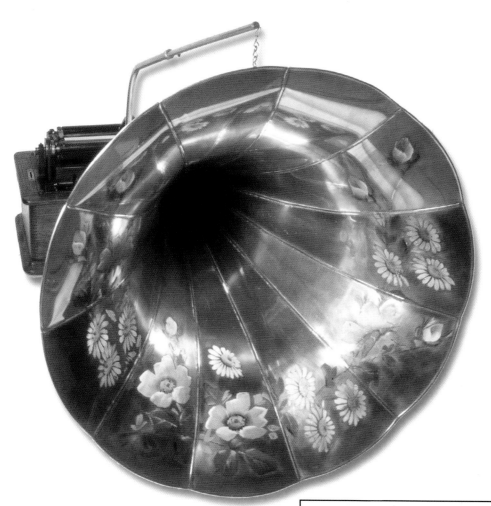

3-142. This unmarked brass flower horn features a most unusual and spectacular assortment of blossoms. Measures 31" long, 22 3/4" in diameter. *Courtesy of David and Lerria Rosamond (Value code: VR)*

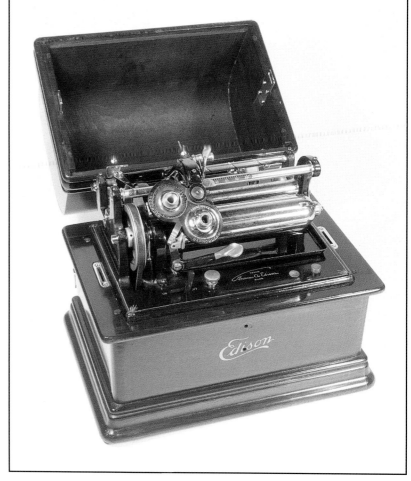

3-143. The Edison "Business Phonograph" appeared in July 1905, at first utilizing oak cabinets similar to, but slightly larger than the "Standard." This example employs 1909-style base molding on — of all things — a *mahogany* cabinet. Someone must have had a pretty plush office for which he needed to put in a special order. Wooden cabinet "Business Phonographs" were soon replaced by the metal-cased "Ediphone." *Courtesy of Richard Goodin (Value code: VR)*

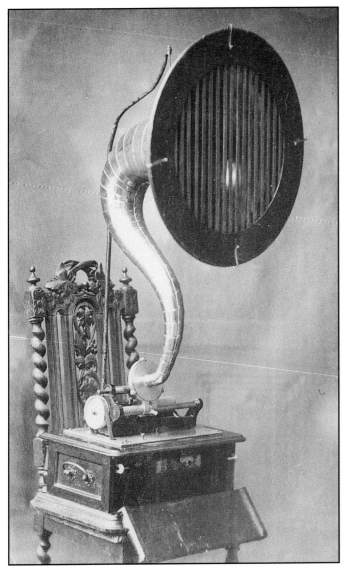

3-144. A period photograph showing a modified Type "BC" Graphophone. This model was introduced in 1905 as a "Loud-Speaking Graphophone," employing a Higham mechanical amplifier and large 4" diaphragm. Manufactured through 1909, early production models had nickel-plated carriages (as seen here), while later carriages were painted black and accented in gold. This peculiar example had a custom-made horn (evidently inspired by the 1909 Edison "Cygnet" horn) fitted to it, supported by an unsophisticated crane. Note the grille in the mouth of the horn attached by four large clips. The photographer made a point of keeping the cabinet door open, revealing what appears to be an electric motor inside the dark interior. *Courtesy of the Charles Hummel collections (Value code: K)*

3-146. The French were fond of phonograph "Salons" and tokens from such establishments can be found in various forms. *Courtesy of Stacey Murdock (Value code: K)*

3-147. The Germans were not immune to the charms of coin/token activated recorded music, and Pathé aggressively pursued this market throughout Europe. *Courtesy of Stacey Murdock (Value code: K)*

3-148. The obverse of the German Pathé token features the company's discus thrower. *Courtesy of Stacey Murdock (Value code: K)*

3-149. Looking at only one side of this token raises the question if it was meant for redeemable services at the Stofflet Phono Shops, or if it was simply an advertising piece. In either case, "Radio" had taken top billing, placing this artifact in the late 1920s. *Courtesy of Stacey Murdock (Value code: K)*

3-145. Coin-operated phonographs sprouted in 1889 and have continued to exist through periods of greater or lesser popularity. The 1890s and early 1900s were especially robust for coin-operated music, and oftentimes the devices were adjusted to accept tokens rather than coins. This rather enigmatic token marked "BB" was "Good For One Free Play On Phonograph." *Courtesy of Stacey Murdock (Value code: K)*

3-150. A glance at the obverse of the Stofflet token shows that it was clearly a "discount coupon" for the firm. During the 1920s, coin-operated phonographs experienced a lull, but the Great Depression would revive the popularity of dropping a nickel in the slot. *Courtesy of Stacey Murdock (Value code: K)*

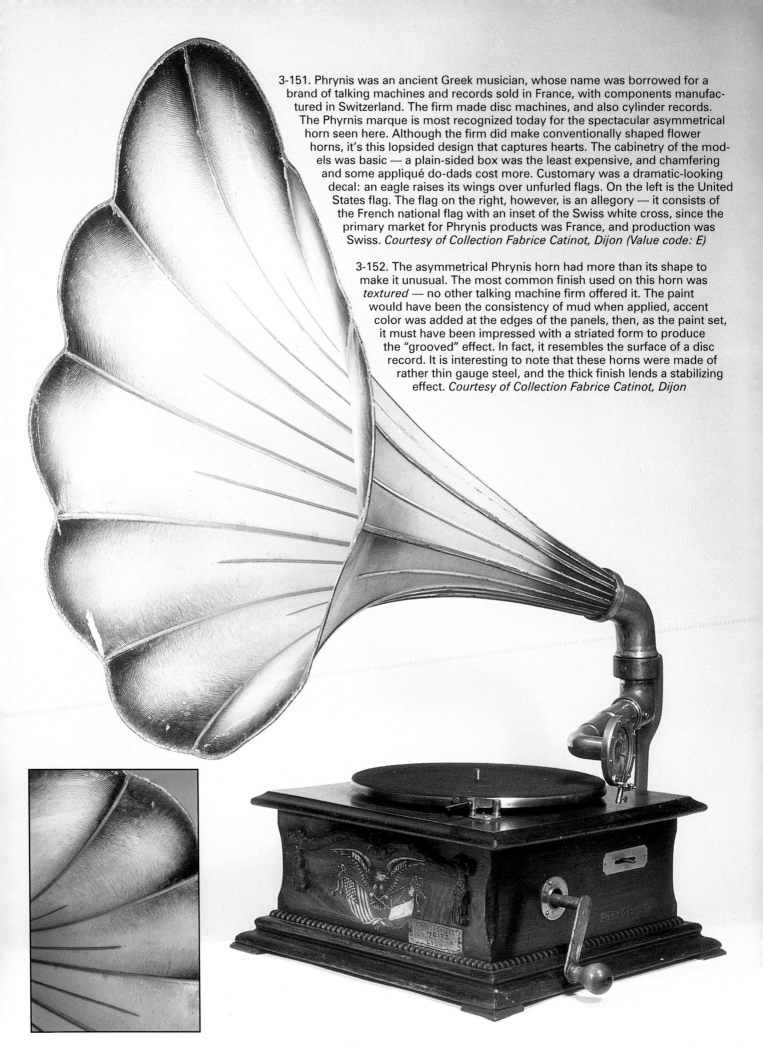

3-151. Phrynis was an ancient Greek musician, whose name was borrowed for a brand of talking machines and records sold in France, with components manufactured in Switzerland. The firm made disc machines, and also cylinder records. The Phyrnis marque is most recognized today for the spectacular asymmetrical horn seen here. Although the firm did make conventionally shaped flower horns, it's this lopsided design that captures hearts. The cabinetry of the models was basic — a plain-sided box was the least expensive, and chamfering and some appliqué do-dads cost more. Customary was a dramatic-looking decal: an eagle raises its wings over unfurled flags. On the left is the United States flag. The flag on the right, however, is an allegory — it consists of the French national flag with an inset of the Swiss white cross, since the primary market for Phrynis products was France, and production was Swiss. *Courtesy of Collection Fabrice Catinot, Dijon (Value code: E)*

3-152. The asymmetrical Phrynis horn had more than its shape to make it unusual. The most common finish used on this horn was *textured* — no other talking machine firm offered it. The paint would have been the consistency of mud when applied, accent color was added at the edges of the panels, then, as the paint set, it must have been impressed with a striated form to produce the "grooved" effect. In fact, it resembles the surface of a disc record. It is interesting to note that these horns were made of rather thin gauge steel, and the thick finish lends a stabilizing effect. *Courtesy of Collection Fabrice Catinot, Dijon*

3-153. Phrynis deserved the award for the most imaginative disc talking machines in the French market. This posh model featured fancy marquetry and a huge brass horn. *(Value code: VR)*

3-154. The terminology "Salon" was employed by the Pathé company in two very different capacities. Firstly, the term referred to a line of uniquely-sized cylinder records, approximately 3 1/2" in diameter, when they were marketed in Great Britain. (English speakers today are inclined to call these records by the "Salon" name, but the term for them when marketed outside the UK was "Inter," meaning intermediate. We embrace the Inter nomenclature in our books because more records were sold under that designation.) The second use of "Salon" was in reference to a category of Pathéphone (disc talking machine) in which the mechanical works and horn were enclosed in a finely-finished, freestanding wooden closet. Here we see the Salon "No. 11," in "Empire" style, measuring a formidable 23" wide, 24 1/2" deep and 66" high. *Courtesy of Phonogalerie, Paris (Value code: VR)*

3-155. The interior of the "No. 11" reveals that it is merely a conventional Pathéphone enclosed in an elegant cupboard. *Courtesy of Phonogalerie, Paris*

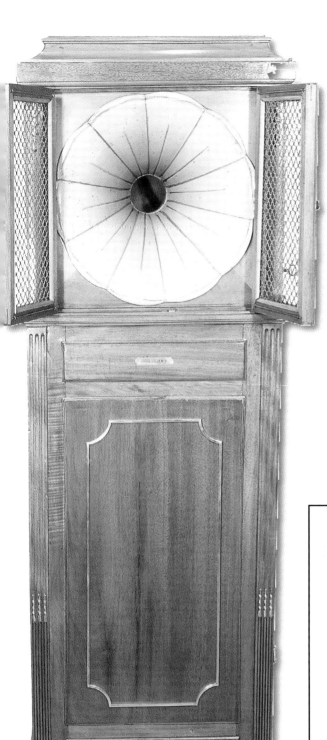

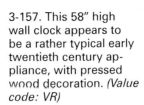

3-157. This 58" high wall clock appears to be a rather typical early twentieth century appliance, with pressed wood decoration. *(Value code: VR)*

3-156. Another Pathé Salon instrument, this time No. 9, in Louis XVI style, measuring 21 1/2" wide, 21" deep and 67" high. *Courtesy of Phonogalerie, Paris (Value code: VR)*

3-158. Within, the basic mechanism of a Pathé "La Jeunesse" (youth), circa 1909, will play a small-diameter Pathé disc when the "alarm" is triggered. The concept of replacing the customary ringer with music or song persisted for years (see *The Talking Machine, an Illustrated Compendium*, Fabrizio & Paul, figure 7-22).

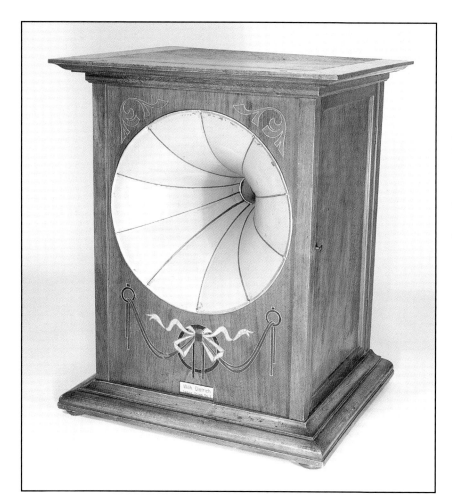

3-159. One of the smaller instruments in the Pathé "Salon"/"Concert" series. Both terms were used in France to denote Pathéphones enclosed in wooden cabinets. In Great Britain, only "Concert" seems to have been used in reference to machines of this type. The cabinet once had a decorative header, now missing. *Courtesy of Garry James (Value code: F)*

3-160. Within the Pathéphone are conventional works discreetly hidden from view. *Courtesy of Garry James*

3-161. The 1910 Pathé catalog lists this Pathéphone Modèle No. "8," in a walnut cabinet with carved panels, at 110 FF. Pathé horns, over the many years they were offered, were available in a variety of opaque and translucent finishes, and a wide variety of colors including white, black, light blue, dark blue, pink, red, light green, and dark green. *Courtesy of Michael D. Wallace (Value code: F)*

3-162. The No. "12" was about as fancy as an external-horn Pathéphone table model got. In the 1910 catalog, it was top-of-the-line at a sobering 175 FF. Yet, as the next illustration will show, Pathé reserved an even grander model for the filthy rich. Take special note of the horn — the surface of the steel was invested with an irregular pattern, by a process similar to electro-plating. Translucent gold lacquer accentuated this "galvanized" effect, creating the appearance of ripples or ice crystals. In fact, the same sort of bold finish was available from American horn manufacturers Hawthorne & Sheble, under the name "Crystal." *(Value code, with this special horn: VR)*

3-163. The Pathéphone No. "14" was the French equivalent of the Victor "VI." Stylish mahogany cabinet, gold-plated hardware, immense white horn with decoration on each panel — this instrument had all the extras, where money was no object. It's an impossible rarity today, especially in such pristine condition. *Courtesy of Jalal and Charlotte Aro (Value code: VR)*

3-164. The near-constant litigation between the Columbia Phonograph/American Graphophone companies and the Victor Talking Machine Company throughout the first decade of the twentieth century resulted in various cross-licensing agreements. By these agreements, Columbia could market machines with tapering tone arms protected by a Victor patent, and Victor in turn could market discs originally recorded in wax, a process covered by a Columbia patent. A point of contention, however, appears to have been Victor's "Sound-Modifying Doors" as employed on its ubiquitous Victrolas. When Columbia introduced its "Elite" in 1909, it was equipped with two simple, Victor-like doors to control volume. In short order (probably as a result of threats of legal action by Victor), the "Elite" was modified with baffles (as shown here). These baffles were also a short-lived feature, since the "Elite" was rapidly equipped with adjustable louvers as seen in later Columbia Grafonolas. *Courtesy of Harvey P. Kravitz (Value code: H)*

3-165. Was it Christmas that inspired the design team at Compagnie Française du Gramophone to create this one — or the traditional décor of a Chinese restaurant? It's an uncataloged model, equipped with "No. 24B" brass flower horn. Note the "La Voix de Son Mâitre" logo on the front of the cabinet. *Courtesy of Jalal and Charlotte Aro (Value code: VR)*

3-166. From the 1909 Spanish catalog of Compañia Francesa del Gramophone came the "Gramophone Monarch núm. 7" with ram's head decoration. The mechanical parts and hardware were manufactured by Victor. It is interesting to note, however, that Victor customarily did not offer all-brass straight horns such as this one domestically. In the United States, Victor straight horns were brass-bell, steel body. European Gramophones, however, followed the model of Zonophone, which sold all-brass straight horns the world over. *Courtesy of Jalal and Charlotte Aro (Value code: F)*

3-167. The Victor/Gramophone group of associated companies sold products that were often specifically designed for individual ethnic markets — for instance, this awesome-looking Gramophone from the Spanish catalog. At first, it appears utterly different than other Victor/Gramophone instruments. Albeit the cabinet is specialized, note familiar Gramophone elements, such as the dial speed control, the brake, the crank and escutcheon, the back-bracket, arm and soundbox, elbow and brass paneled horn (sold in the United States as the No. "24B"). The motor, too, is typical Victor/Gramophone, and all these parts were manufactured in Camden, New Jersey, and shipped abroad. A similar model, "Gramophone Monarch núm. 13," appeared in a 1909 catalog. *Courtesy of Phonogalerie, Paris (Value code, this rare variation: VR)*

3-168. On the rear panel of the Spanish Gramophone, the familiar "His Master's Voice" logo. Two trade marks were associated with British and European Gramophones. During the early part of the twentieth century, the "Recording Angel" trade mark (see figure 3-93) was more prominent than depictions of the dog Nipper. At the end of the first decade, "His Master's Voice," in a number of languages, took precedence. Within the Victor/Gramophone hierarchy, products seemed to flow from Camden to Great Britain, and then on to Europe. Compagnie Française du Gramophone (the French branch) was responsible for Spain. We see here the abbreviation for, "Compañia Francesa del Gramophone." *Courtesy of Phonogalerie, Paris*

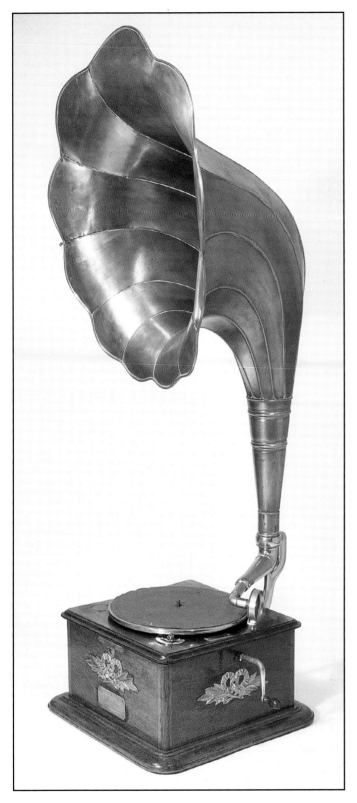

3-170. The brass ID of the Fontanophone lists the name of Mr. A. Fantanon, "former soloist," who manufactured the machines in Paris. *Courtesy of Phonogalerie, Paris*

3-169. The concept behind the Fontanophone (circa 1910) was to achieve the greatest possible acoustic amplification. Therefore, the horns with which the firm's otherwise ordinary disc talking machines were fitted were very large and rather oddly shaped. Another French brand, the Aérophone (see figures 3-178 and 3-179) employed a similarly tall, though less attenuated horn. *Courtesy of Phonogalerie, Paris (Value code: VR)*

3-171. A striking German coin-op with appliqué busts of Beethoven, and concentrically-formed brass horn. Many German coin-ops embraced this same design: An apparatus emerging from the top of the cabinet near the pivot point of the tone arm to control its movement, a cabinet raised on "feet," a large, eye-catching horn. It's difficult to identify specific brands and models, since they were very seldom marked and produced in such abundance. *Courtesy of Phonogalerie, Paris (Value code: E)*

3-172. A German talking machine with three horns. During the acoustic phonograph era, there were a number of double or triple horn technologies. In this instance, the air pressure is split between the three horns, not achieving more volume, but creating multi-directional sound. In other words, the instrument looks a lot more impressive than it really is. Note the nickel-plated horns with red-painted interiors, a style popular in Europe yet virtually unknown in the United States. *Courtesy of Jalal and Charlotte Aro (Value code: VR)*.

3-173. An astonishingly bright German talking machine. The back support bracket, reminiscent of a ship's figure-head, is the only one we've seen. *Courtesy Musée de l'Aventure du Son, St. Fargeau (Value code: VR)*.

3-174. Curvilinear is the word for the Columbia Disc Graphophone Type "BY" ("Improved Imperial"), here shown with a matching record cabinet, circa 1910. *Courtesy of Jerry Blais (Value code, with matching cabinet: VR)*

3-176. A close-up of the graceful female figure. From the examples of this machine that have been observed, the horn was commonly finished in translucent green. *Courtesy of Phonogalerie, Paris*

3-177. Special records such as this were used in the Concert coin-op. They were marked "not for sale." A special run-in groove allowed the mechanism to get up to speed before the selection began, and a uniquely-designed run-out groove triggered the mechanism to stop. Most coin-op disc instruments employed a simpler system, whereby they were timed to play for a certain pre-set period. *Courtesy of Phonogalerie, Paris*

3-175. The Pathé Concert coin-op of 1910 brings to mind a three-dimensional expression of *belle epoch* Czech artist Alphonse Mucha's work. This vertical cut disc playing machine was available in either electric or spring-driven versions. In this instance, the winding crank (not shown) fits into the hole located under the figure's legs. Measures 24 1/2" wide, 27" deep and 67" high. *Courtesy of Phonogalerie, Paris (Value code: VR)*

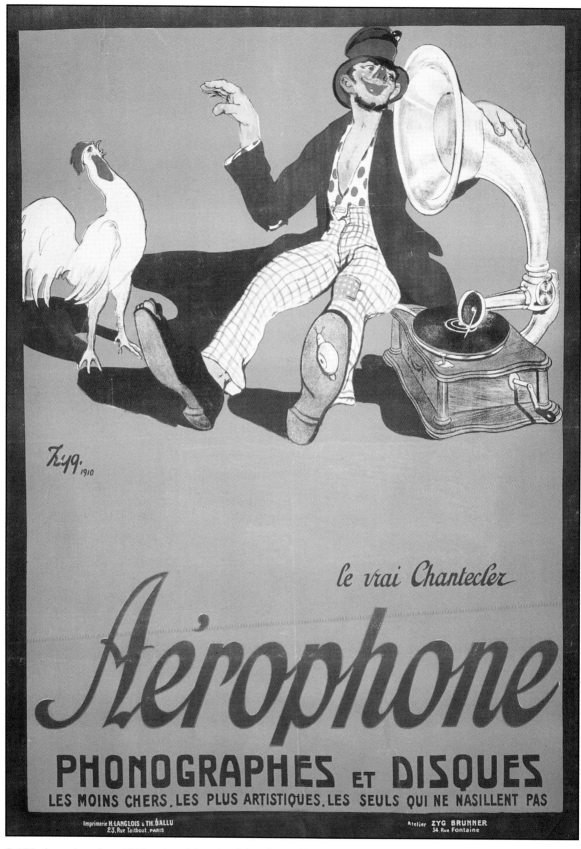

3-178. A poster, circa 1910, advertising the Aérophone. The *clochard*, or hobo, dismisses the vocal efforts of a rooster (at least it's not a red Pathé *coq*) and declares the talking machine "the true Chantecler (the rooster protagonist from an eponymous play by Rostand, well known at the time). Furthermore, we learn that the Aérophone phonographs and records are "less expensive, more artistic, and the only ones that don't sound nasal." 44 1/2" x 61". *Courtesy of Phonogalerie, Paris (Value code: VR).*

132

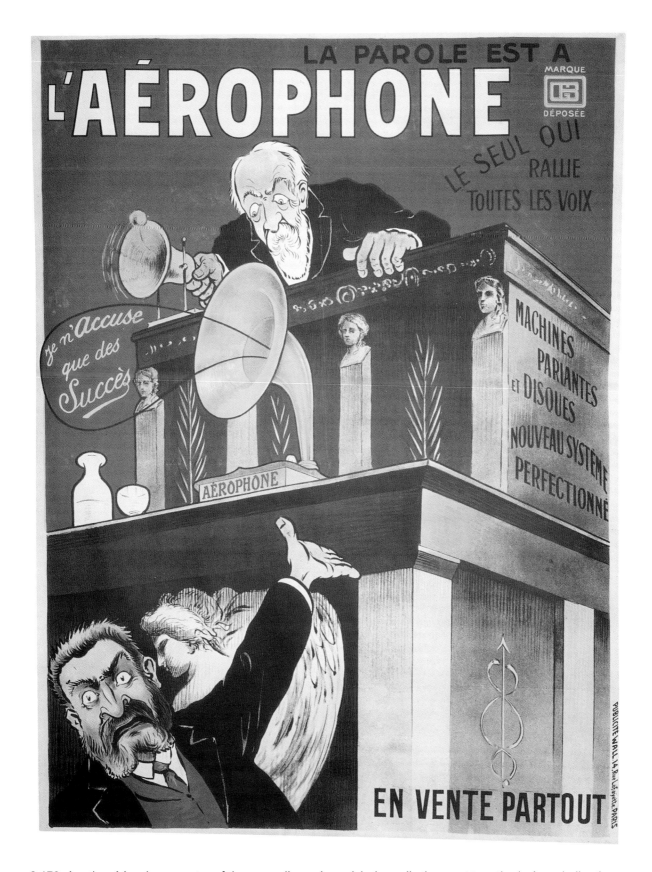

3-179. Another Aérophone poster of the same dimensions. A judge calls the court to order (using a bell rather than a gavel), as the talking machine announces "I accuse only of success." The reference to the court martial of Captain Dreyfus for espionage in 1894, and Zola's famous allegations of a frame-up in an open letter, nicknamed *J'accuse!*, would have resonated (pun intended) with the French viewers of this poster. Dreyfus had only recently (1906) been cleared of all charges. *Courtesy of Phonogalerie, Paris (Value code: VR)*

3-180. On this poster, a scholar is startled to hear a machine declare, "Don't write any more." He drops the tools of his pedagogy at the thought of such a disturbance of the status quo. "Why?" continues the advertising copy, "Now send your correspondence by the Phonopostal [recording/playback machine] and the Sonorine [recordable post card]." Circa 1910, 44 1/2" x 61", (see *Antique Phonograph Accessories and Contraptions*, Fabrizio & Paul, figure 3-115) *Courtesy of Phonogalerie, Paris (Value code: VR)*.

3-181. A German coin-op in the Mammut vein, with a characteristically Teutonic oak cabinet. Although instruments of this sort always employed large horns, this particular one is fitted with a "Symphonista" attachment, one of a line of extra-big horns that could be appended to many German (and Pathé) talking machines using common back-bracket fittings (see *Antique Phonograph Accessories and Contraptions*, Fabrizio & Paul, figure 1-107). Measures 71" to the top of the horn. *Courtesy of Larry and Myra Karp (Value code: D)*

3-182. Toward the end of the twentieth century's first decade, a few inventors developed automatic stops for disc talking machines. To the uninitiated, the need for such a device might seem questionable, but owners of disc playing machines soon discovered that the turntable acted as a flywheel, causing the motor to continue even after the mainspring was relaxed, with the possibility of flexing and unhooking the mainspring from the winding shaft. To avoid the attentions of a repairman, this Simplex "Automatic Start and Stop Device" solved the problem for only $3.00. (See also the frontispiece of *Antique Phonograph Gadgets, Gizmos, and Gimmicks*, by Fabrizio & Paul.) *Courtesy of Ernest Carl Allen (Value code: K)*

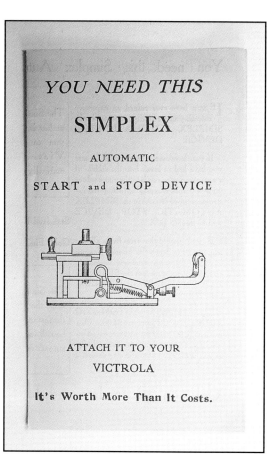

3-183. A view of the actual Simplex Automatic Start and Stop Device on a Victor "III." A small brass wheel is mounted at the end of a spring-loaded arm to the right of the tone arm. By pushing the tone arm against this wheel, the Simplex's adjustable arm is depressed and cocked, releasing the brake at the rear of the turn table. As the record plays, the tone arm eventually makes contact with a sliding boss to the left. A spring-loaded copper plunger within the assembly is activated, driving the Simplex's arm downward and causing a copper wedge to push the brake against the turntable. Despite its complexity, the device works surprisingly well. *(Value code: J)*

3-184. A rear view of the Simplex shows details of its adjustments, as well as its various components.

3-185. Arthur J. O'Neill was not a man to let grass grow under his feet. He ran two talking machine firms, the O'Neill-James Company and the Aretino Company, in Chicago during the first decade of the twentieth century. What was unknown until recently is that he also sold vacuum cleaners under the Aretino mark. *Courtesy of Steve Hosier (Value code: VR)*

3-186. The cleaner was decorated in colors associated with Aretino disc talking machines: green, like Aretino horns, and red — in fact both devices used the same red banner emblazoned "The Aretino Machine." *Courtesy of Steve Hosier*

3-187. From the looks of this lady's outfit, we'd say the Aretino vacuum cleaner would date from around 1910. That would put it near the end of the brand's life in the talking machine field, when Mr. O'Neill was likely seeking ways to diversify. In a letter to a prospective dealer dated March 22, 1911, Arthur J. O'Neill announced, "…our new BUSY BEE LIGHTNING VACUUM CLEANER. The Workmanship of the BUSY BEE is first class in every respect… It is built very compact — simple to operate and is the only GUARANTEED Machine at $15.00… Remember, every home is thinking of buying a VACUUM CLEANER and you have the opportunity to get in on the ground floor." "Busy Bee," like "Aretino," was also a brand of talking machines which O'Neill sold. Whether "Busy Bee" vacuum cleaners replaced "Aretino" or vice versa is not known. *Courtesy of Steve Hosier*

3-188. Typical of the German coin-ops produced in abundance around 1910 is this "Parlophone." In the better models, the arm was automatically controlled to reset after the record finished playing. *(Value code: E)*

3-190. For precision purposes, such as sound synchronization of motion pictures, spring motors were considered to be too unreliable. Weight drive was the most even, but an electric motor, if mechanically governed, could perform almost as well. Here we see the workings of a Pathé talking machine mechanism very similar to the one seen in the previous illustration, except that it is driven by an electric motor. A great deal of effort has been made to provide the steadiest performance possible. *Courtesy of Jalal and Charlotte Aro (Value code: VR).*

3-189. Weight driven motors were applied to talking machine mechanisms in several instances. The London Stereoscopic Company produced a weight-driven Tinfoil Phonograph in the early 1880s. French inventor Henri Lioret employed the concept in some of his Lioretgraphs. Here, the notion was utilized by Pathé, but in what precise capacity is not known. It might have had some connection with sound for films. Pathé started in the motion picture business at the turn of the twentieth century, at about the same time its phonograph enterprise began to blossom. The first two decades of the century saw numerous efforts by many individuals to synchronize sound with picture. In the case of this particular apparatus, the descending weights afforded an especially steady and long-lasting source of power, such as would be required for coordinating a recording with a moving image. The reproduction system was vertical cut — and note the unusual aluminum and copper horn. The overall design was clearly industrial, with a stout-timbered stand, and heavy cast iron weights that bear the Pathé name. *Courtesy Musée de l'Aventure du Son, St. Fargeau (Value code: VR)*

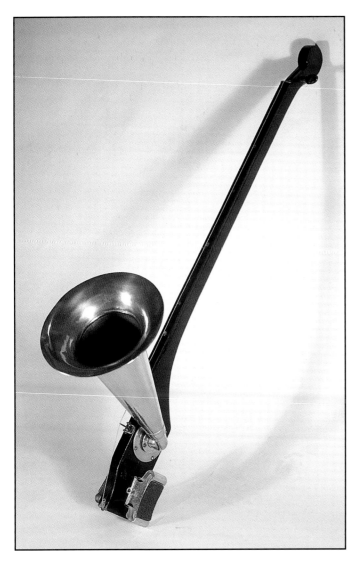

3-191. "Diaphragm-amplified" stringed instruments enjoyed a vogue at the turn of the twentieth century. These devices employed a phonograph-like system, whereby the vibrations of the strings were communicated to a diaphragm, like that of a talking machine soundbox, and augmented through a horn. Modern collectors often associate these instruments exclusively with early recording, since conventional stringed instruments failed to be captured until "diaphragm" apparatus arrived. It is interesting to note, however, that the initial exploitation of these contrivances was by music hall performers. Shown here is a Stroh, one-string or "Japanese" fiddle, a very popular item at the time. *(Value code: I)*

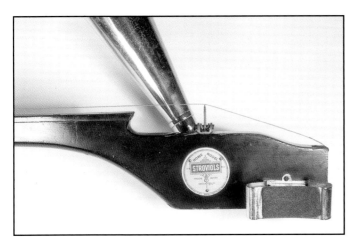

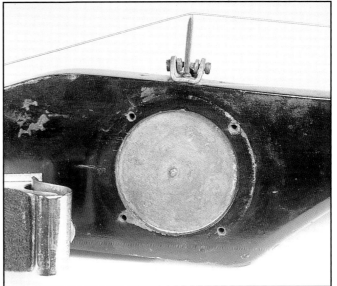

3-193. The Stroh diaphragm, with protective plate removed. Augustus Stroh spent years experimenting with the response of diaphragms to sound vibrations.

3-194. This advertisement in a 1910 issue of the *Talking Machine World* illustrated the Stroh "One String Fiddle" and "Violin," and reminded dealers that Stroh instruments were available in the United States from the Oliver Ditson Company. Although the Stroh family name is exploited here, Evans usually suggested it in an indirect way — his trade mark, "Stroviols" (registered

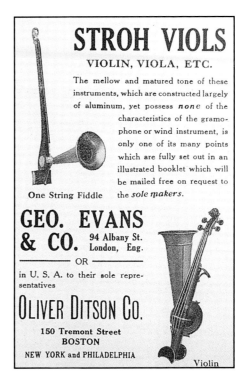

STROH VIOLS
VIOLIN, VIOLA, ETC.

The mellow and matured tone of these instruments, which are constructed largely of aluminum, yet possess *none* of the characteristics of the gramophone or wind instrument, is only one of its many points which are fully set out in an illustrated booklet which will be mailed free on request to the *sole makers*.

One String Fiddle

GEO. EVANS & CO. 94 Albany St. London, Eng.

OR

in U. S. A. to their sole representatives

OLIVER DITSON CO.
150 Tremont Street
BOSTON
NEW YORK and PHILADELPHIA

Violin

in 1910), omitted the "h" and personalized the business he had acquired.

3-192. A close-up of the Stroh identification. Naturalized British subject Augustus Stroh, an acoustic science pioneer, had developed the idea from his sound experiments of the 1880s. In 1899 he patented a "diaphragm-amplified" violin in England. He subsequently began manufacturing it in collaboration with his son, Augustus Charles, known as Charles. To illustrate further how the story of the Stroh instruments was intertwined with that of the talking machine, Stroh's entire production was taken over in 1906 by the Russell Hunting Record Company, Limited. The business failed, however, and was assumed by George John William Evans in 1909.

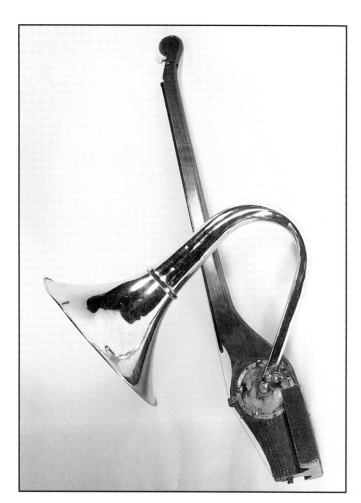

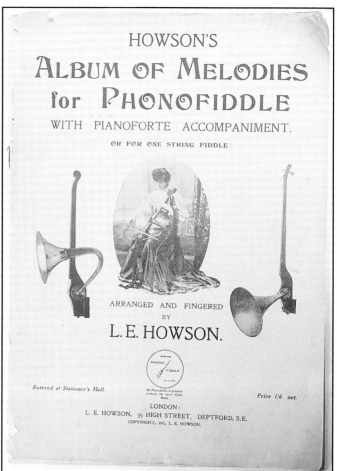

3-195. Howson's "Phono-Fiddle" was an instrument similar to Stroh's. Arthur Thomas Howson claimed he invented it in 1886 (though that is difficult to prove), and the device was popularized in vaudeville during the first decade of the twentieth century. A novelty act performed by G.H. Chirgwin (who appeared in black-face, with a white diamond painted around his right eye) featured Howson's "Jap fiddle." Chirgwin had previously performed with ordinary stringed instruments. *(Value code: I)*

3-197. Although limited to one string, the "Phono-Fiddle" was versatile enough to have an album of music arranged for it by L.E. Howson (Laura Ellen, Arthur's wife). This 1911 publication included works by Handel, Schubert and Gounod, with piano accompaniment. *(Value code: K)*

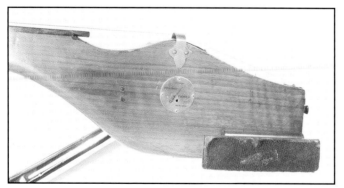

3-196. Howson's trade mark. From *The Hippodrome* (a music hall periodical), Summer Number, London, 1911: "If you have seen Mr. G.H. Chirgwin, the inimitable White eye'd Kaffir, you have not only heard the 'concert' phono-fiddle, but you have heard it almost speak… Mr. Howson… is the father of the concert phono-fiddle, and it is due entirely to his exertions, that this new musical instrument… has attained its present popularity. When an invention comes on the market there are always, unfortunately, a large number ready to imitate it, and although 'imitation is the sincerest form of flattery,' it is none the less annoying to find that an instrument, no more a real phono-fiddle than a banjo is a violin, has been unwittingly purchased…" That's a heavy-handed reference to rival George Evans — or might people be buying Howson's device thinking it was a "Stroviol"?

3-198. A British cabinet card celebrating Fred Wildon. "The Original Phono-Fiddle Soloist." We would like someday to hear a recording of a virtuoso such as Mr. Wildon playing the one-string instrument — the popularity it enjoyed during this period is difficult to imagine. *Courtesy of Bill Boruff (Value code: K)*

4-1. The Morel Orphée was designed along the lines of the Pathé "Jeunesse" (Youth), having a high, narrow profile, with a shallow internal-horn, playing vertical-cut records. In fact, the reproducer was certainly manufactured by Pathé, and probably the other mechanical parts as well. The cabinet employed the prosaic Orpheus/lyre motif seen in a wide variety of other talking machines, from the humble Puck to the imposing first version of the Edison Amberola "I-A." *Courtesy of Jalal and Charlotte Aro (Value code: VR)*

4-2. Hardly another talking machine company worked as tirelessly as Pathé to produce ingenious products in a dizzying array of variations. One only need examine their record catalog, especially the disc portion of it, to observe an astonishingly huge repertoire presented in a plethora of diameters. Why would any firm put such effort into so many individual products, many of which must have sold in modest numbers? It could only have been true dedication to the art and science of recorded sound. Which is not to suggest that Pathé was not successful — it was. However, it probably could have achieved equal profits with half the number of records and phonographs. With this in mind, we look upon the superbly conceived and executed Pathé-graphe. The enameled metal lid which covers the mechanism brings to mind the case of an antique typewriter — and appropriately so, since both appliances have an educational purpose. *Courtesy of Phonogalerie, Paris (Value code: F)*

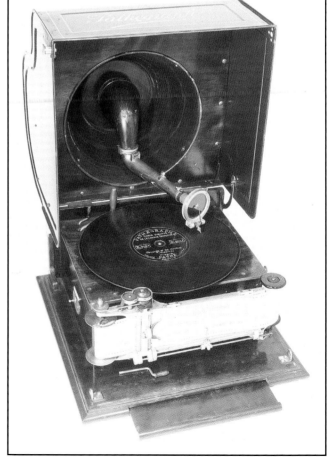

4-3. Opened, the components viewed in the foregoing figure may be seen in place and ready for language teaching. The record contains the spoken lesson. The *bande*, or lesson strip, contains the written words. The sound is delivered by the "reflex" system, which Pathé favored during the 1910s. Thus, with an utter sense of precision, the client is ready to learn his first three words in German… *Courtesy of Phonogalerie, Paris*

4-4. A circa 1912 advertising plaque, measuring 14 1/2" x 20 1/2", depicting the brothers Pathé, Emile and Charles, marching toward "the conquest of the world." An internal-horn Pathéphone, a cinema projector and a rampant rooster are the only allies they need to get the job done — coupled with the traditional tenacity and pluck of both the French and the bossy barnyard bird (though "pluck" might be an unfortunate choice of words in reference to a chicken). Signed "A. Barrère" (French artist and caricaturist Adrien Barrère, 1877-1931). *Courtesy of Jalal and Charlotte Aro (Value code: VR)*

4-5. The concept of running two reproducing points close together in the same record groove to achieve "double the sound" was exploited in a number of talking machines. In the cylinder mode, there was the "Polyphone" of 1898. Several disc instruments were based on this idea, including the Pathéphone seen here — a kind of "Jour et Nuit" on steroids! The "Jour et Nuit" (Day and Night) was a popular model, adaptable for playing with full volume through an external-horn (to overcome the extraneous noises of the day), or softly through a rather small internal-horn (at night, when the neighbors were trying to sleep). In this double version, the volume is increased by a pair of horns (either the external or the internal set). The playing unit on the right functions conventionally, as would an ordinary "Jour et Nuit." The unit on the left requires a specially shaped sapphire stylus for the reproducer, since it meets the record in the opposite position. The 1912-13 British Pathé catalog offered a simplified, internal-horn-only variation, called the "Duplex" Pathéphone. It was merely two hornless Pathéphones side-by-side, each with its own turntable, arm and soundbox. The purpose was to allow a "seamless" flow of music by switching from one turntable to the other. *Courtesy Musée de l'Aventure du Son, St. Fargeau (Value code: VR)*

143

4-6. The violin is delicate and refined, and talking machine marketers wanted to associate their products with it. The Vitaphone of 1912 was described thus, "An elementary knowledge of tone appreciates the fact that the most natural and musical tones are produced through the medium of WOOD rather than metal. The violin, the flute, the organ pipe, the 'cello are but common types which prove the soft, mellow tones of wood which improve as time goes on. The Vitaphone is constructed along this distinctive line…" The Heywood-Wakefield "Perfek'tone" wicker cabinet talking machine, circa 1920, was said to have an internal-horn "…composed of a matrix of wood and fabric having a peculiar vibratory action of its own, and gives a fullness and sweetness of tone which can be compared to a rare old violin." Now picture this — why settle for comparing a part of your machine to a fine violin, when you could use a violin itself to amplify the music? *Voilà!* "Le Palmodian." Frenchman Henri-Olympe Buffet patented the device, and he manufactured it in France in 1912. There were two sizes, and each employed a violin specially created for the purpose of record reproduction. An elaborate mounting transferred the vibrations from the needle to the body of the instrument, and the results were quite successful. We're not sure if the folks at Klingsor influenced the inventor to insist the strings be specifically tuned. "Le Palmodian" was sold under the motto, "Le violon qui chante" (the violin that sings). The mystery of the name may be found in Latin. Buffet combined the root *pal* (to strike), as in *palpito/palpitare* (to throb), with *modus* (rhythm in music), and added the ending *ian* to indicate a person or object possessing or admiring this quality, i.e.: throbbing or pulsing rhythm in music. *Courtesy Musée de l'Aventure du Son, St. Fargeau (Value code: VR)*

4-7. The U-S Phonograph Company of Cleveland, Ohio, offered some fascinating cylinder technology during its short life (1910-1913). The U-S line was announced to the trade in May 1910, and Edison almost immediately unleashed a legal barrage against the new firm. U-S Everlasting cylinders were offered in monthly releases through April 1913. For the following six months, new titles appeared irregularly; finally sputtering out after October 1913 as the company folded beneath the costs of lengthy litigation, even though it had won against Edison. Shown is the top of the firm's external-horn models, the "Opera" ($65.00 in oak; $75.00 in mahogany as shown). The massive carriage carries two reproducers, one for two-minute cylinders; the other for four-minute. The lid handle is missing from this example. *Courtesy of the Domenic DiBernardo collection Value code: VR)*

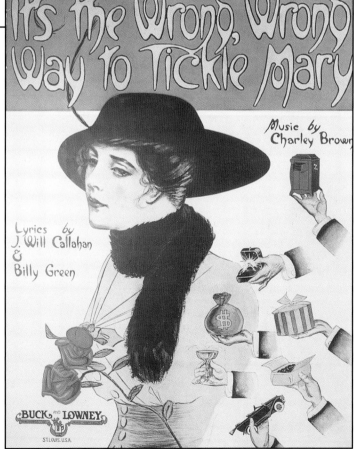

4-8. "It's a Long Way to Tipperary" was written in 1912, and enjoyed two periods of popularity — after its initial success as a derogatory Irish novelty, it was taken up as a sort of anthem by soldiers in the Great War, who largely ignored the Irish-bashing verses and marched along to the jaunty refrain. Yet, what's a popular song without a parody? Parodies, of course, presuppose a familiarity with the original — hence, the title of this song plays off the words of the refrain (… long, long way). It's interesting to note that among the blandishments being offered to a rather chilly Mary, besides chocolate, jewelry and a motor car, was a Victrola! *Courtesy of Bradley Kuiper (Value code: K)*

4-9. When Thomas A. Edison, Incorporated introduced the Edison Disc Phonograph in 1912-1913, the emphasis was on high-grade instruments and highbrow music. In spite of the fact that Victor was offering a Victrola for as little as $15.00, the least expensive Edison Disc Phonograph was $60.00, and prices skyrocketed from there. Of the new line, one model — the "A-250" — borrowed its cabinet design from Edison's cylinder playing Amberola "IB." Bowing to the demand for instruments finished in rare and expensive Circassian walnut, Edison offered the same cabinet in this luxurious wood for $300.00, and thus was the "A-300" born. *Courtesy of Jon and Jackie Kelm (Value code: VR)*

4-10. The interior of the Edison "A-300" is a sumptuous amalgam of beautiful grain, brown enamel, and gold plating. Note the protective guard preventing the handle which engages the reproducer from being accidentally initiated and dropping the stylus on the record. *Courtesy of Jon and Jackie Kelm*

4-11. Beginning in 1912 and repeated annually, the employees (and their families) of the various companies that comprised Thomas A. Edison, Incorporated participated in sponsored games and activities called the "Edison Field Day." Typical track and field events such as the 100, 220, and 440 yard dash, shot put, triple jump, and high jump were rounded out with an obstacle race, sack race, three-legged race, stake driving, wheelbarrow race, fat man's race, shadow race ("for the narrow fellows"), special events for ladies, and baseball. The competitions were well-organized and enthusiastically contended, as attested by this trophy from the 1919 Field Day. Employee J.J. Kennedy won laurels for himself and the Phonograph Works by amassing the most points in a variety of events. *Courtesy of John P. Andolina Jr., The Early Sound Man (Value code: VR)*

4-12. This photo was taken at the first Edison Field Day on July 16, 1912, showing the opening pitch of the baseball game. The illustrious pitcher later inscribed the image and autographed it. *Courtesy of Nina and Dave Heitz*

4-13. A close-up of the previous photo shows a steely-eyed Thomas Edison following the track of his pitch. Sandy Koufax, eat your heart out! It's a marvelous candid image of the great inventor, and suggests the surprising vitality of the 65 year old Edison. *Courtesy of Nina and Dave Heitz*

4-14. Pathé was an extremely inventive firm; it designed and produced talking machines with a greater variety of acoustic amplifying systems than any of its rivals in the first three decades of the twentieth century. Pathé introduced a popular series of "Reflex" instruments during the early years of the 1910s. A British Pathé catalog of 1912-13 stated, "...an entirely new method of Musical and Vocal reproduction... the last word in Talking Machines." The smallest of the "Reflex" machines, known as the "Elf" in Great Britain, proved to be the most popular, from our observation. Here we see the largest table model, called the Pathéphone No. "21" on the Continent. *(Value code: H)*

4-15. The No. "21" incorporated a pretty walnut cabinet, quaint grain-painted reflecting dish, and a convenient compartment to the left of the turntable for storing the tone arm and soundbox.

148

4-16. In 1912, the Gramophone Company, Limited sold this "Intermediate Monarch." Although it is illustrated with a straight horn, it is likely to have been equipped with a paneled, or flower, horn. *(Value code: G)*

4-17. On the front of the mahogany cabinet, a colorful depiction of the already world-famous "His Master's Voice" trade mark.

4-18. On the side panel, the distinctive "Gramophone" logo that the British firm was using at the time.

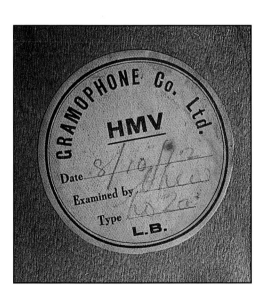

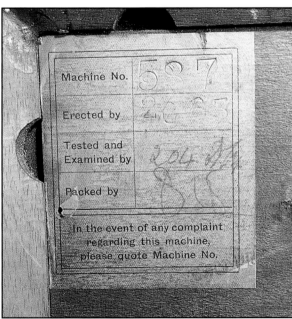

4-19. Rigorous testing and inspection meant that Gramophone products, like their Victor cousins, were among the highest quality goods on the market.

4-20. A Gramophone Company, Limited "Junior Monarch" in oak. The apple green flower horn, seen in Great Britain and Europe, was a relief from Victor's strict imposition of basic black in the American market. *Courtesy of John T. Hoffman, the Phonophile (Value code: G)*

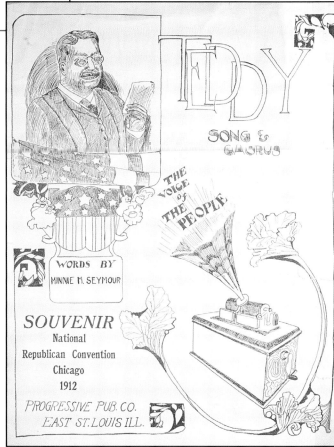

4-21. Teddy Roosevelt was a remarkably popular President during his terms (September 1901–January 1909). His image, adventures and personal affectations inspired advertising slogans (see *Antique Phonograph Gadgets, Gizmos and Gimmicks*, Fabrizio & Paul, figure 5-123), a mechanical bank, the teddy bear, and popular song. Roosevelt was succeeded by William Howard Taft, of whom he approved, but with whom he grew steadily disenchanted. Shaping up was a monumental political struggle for the 1912 Republican Presidential nomination, as Roosevelt determined to pull the rug out from under Taft (not an easy job, literally or metaphorically — Taft was a BIG man). The sheet music shown here was published to exhort and entertain Roosevelt supporters at the Republican National Convention. Taft, however, prevailed, and Roosevelt, smarting from this partisan rebuff, started the "Progressive" party. The ultimate beneficiary of dissension in the Republican ranks was of course Woodrow Wilson, the Democrat, who won the election. The "Progressive Pub. Co." foreshadowed the path Roosevelt would take. *Courtesy of Jerry Blais (Value code: K)*

4-22. Some phonographs were disguised as everyday objects, others were just disguised. It's unlikely most people would recognize this stylized "pavilion" for what it really is. *Courtesy of Stan Stanford (Value code: H)*

4-23. Surprise! The sound travels up the right-most brass tube and is reflected downward by the dome. The words "Bijou" and "Concert" were generic, used repeatedly in reference to various talking machines. *Courtesy of Stan Stanford*

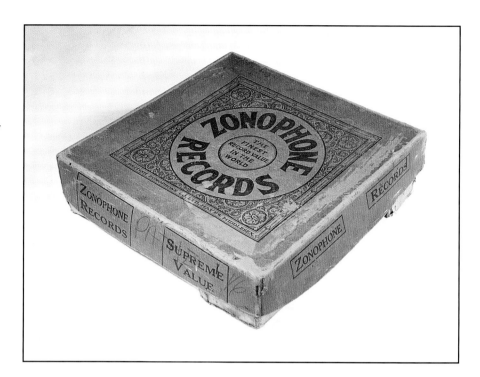

4-24. Circa 1912, British Zonophone 12" diameter discs were shipped in cartons such as this. Note the Hayes, Middlesex, address of the Gramophone Company, Limited, which controlled the Zonophone brand. *Courtesy of Daniel Melvin (Value code: K)*

4-25. Rather than subject salable records to the sun, dust and temperature changes of window displays and other promotional purposes, record companies created "imitation" discs. There is nothing recorded in the grooves of these Victor record simulacra. *Courtesy of Daniel Melvin (Value code: J)*

4-26. Considering the extent of Pathé's success, it's hard to imagine how the firm controlled costs and inventory when it continually offered so many different models. Pathé was the talking machine equivalent of the legendary "New Jersey Diner," which manages to thrive despite an unimaginably huge number of items on the menu. One of the many variations of internal-horn Pathéphones manufactured during the 1912-14 period is shown above. In the English catalogue, it very much resembles the "Onward," a suitably vigorous British epithet. We quote, "The inside Horn is made according to the acoustic laws, which secures for it a reproduction of irreproachable quality. Price: £3/15/0." *Courtesy of Garry James (Value code: H)*

4-27. Furthermore, Pathé had a presence in nearly every part of the globe.

4-28. The "Zig Zag" may resemble a contraption cobbled-up by a hirsute denizen of some mountain cabin with several bedraggled hound dogs on the porch, but it actually has interesting European provenance. It was a premium associated with the purchase of Zig Zag brand cigarette rolling papers and related products. Long before hirsute denizens of the 1960s made Zig Zag an icon of the drug culture, the company was promoting its wares with this vertical-cut talking machine. *Courtesy of Garry James (Value code: G)*

4-29. The horn, really nothing more than a funnel, stored in a space accessed by the front door. The name of the company was derived from the interleaving process by which the cigarette papers were manufactured and packed. *Courtesy of Garry James*

4-30. The Vitaphone Company of Plainfield, New Jersey, maintained a branch in Canada. This Canadian Vitaphone Type "40" displays the company name within the horn located in its lid. The sound-modifying doors (flouting a jealously guarded Victor patent) are cleverly shaped to conform to the lid. *Courtesy of the Domenic DiBernardo collection (Value code: F)*

4-31. With the lid opened, the workings of the Vitaphone Type "40" can be seen. The solid length of "violin wood" through which the Vitaphone's sound vibrations passed was weighted to make the needle exert considerable pressure against the record, and thereby release as much physical energy from the groove as possible. The mica diaphragm of the stationary soundbox was kept under tension, which aided the transference of the vibrations from the arm along a short length of cord. In fact, both laterally and vertically generated vibrations would pass equally well through the wooden arm. This meant that disc records of either type could be played. Upon closing the lid, the ferrule of the diaphragm housing forms a connection with the horn through the hole at the small end. *Courtesy of the Domenic DiBernardo collection*

4-32. This Nipponophone "No. 35" cost 35 Yen, and was described in the catalog as, "Beautifully finished quarter-sawed oak Cabinet, with three-ply hinged top… capable of playing **five records** with one winding. The Horn is flower-shaped, having 9 petals, of brass, nickel-plated, each reinforced." The Nipponophone Company, Limited, of Tokyo, offered a wide variety of external and internal-horn talking machines during the 1910s and 1920s. Under "things you never thought about," Western phonograph enthusiasts might find this catalog quote interesting, "… the wood used in the manufacture of our instruments is thoroughly… kiln dried 3 months previous to the time we place it in the cabinets, we feel satisfied that our instruments will now stand the ravages of any of the trying climatic conditions of the Far Eastern Countries." The company supported "200 Agencies in principal cities throughout the Empire" for the sale of its talking machine, records, and accessories. *(Value code: G)*

4-33. This unusual Columbia Grafonola was based on the popular "Eclipse" model, but with striking added features. The simplified lid design with glass panels is reminiscent of display-only instruments briefly supplied by the rival Victor Talking Machine Company, but the latch at the front of the lid is unique. The enlarged cabinet with storage drawers is equally novel, and one wonders if this was a special order to meet the special needs of some long-ago music lover. *Courtesy of the Charles Hummel collections (Value code: VR)*

4-34. This dainty Columbia suggests the general styling of the company's "Europa" models from 1914. Note that this machine is not labeled "Grafonola," but "Graphophone," as was Columbia's practice for an internal-horn machine with no lid during the 1910s. Despite being marked "Made in U.S.A," it's likely that this model was manufactured in Germany following an arrangement made in April 1914 by Edward N. Burns, vice president of the Columbia Graphophone Company. *Courtesy of Merle Sprinzen (Value code: VR)*

4-35. This view shows the Columbia sound box and speed adjustment. Germany was well known for producing metal goods during the first decade of the twentieth century. A factory in Biersfield, Saxony, manufactured an inexpensive line of talking machines for Columbia until World War I brought the enterprise to an end. Although smaller than the better known Columbia "Europa" models, the design and construction of this example generally follow their pattern. *Courtesy of Merle Sprinzen*

4-36. Except for Canada and South America, where Victor sold external-horn talking machines identical to its domestic line, the company's export business consisted of parts and pieces that were assembled and sold by foreign affiliates. Columbia, however, primarily exported finished external-horn instruments, which in some instances differed from its American offerings. The anomalies largely occurred after the decline of the external-horn market in the United States. The Disc Graphophone seen here, which from its components appears to date circa 1914, is an example of the type of machine Columbia sold in countries such as Great Britain and Italy. A similarly designed instrument was advertised in the Italian publication *Pro Familia* in 1914, called "Grafofono Columbia Tipo [type] 'Prince,' 144 Lire." *Courtesy of Bert Gowans (Value code: F)*

4-37. In the event a customer complained about the quality of the sound coming from his Grafonola, a Columbia dealer could assess the No. "12" or No. "6" soundbox with this "Test Record." The record is numbered "FT-5" which suggests that other test records existed to diagnose various ailments prior to adjustment. A No. "6" soundbox is shown to the left. *Courtesy of the Scott and Denise Corbett collection (Value code: VR)*

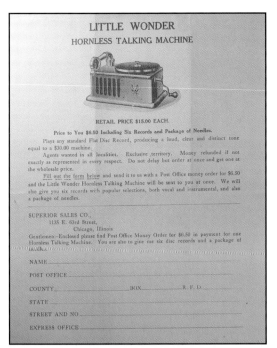

4-38. This advertisement from the Superior Sales Company promoted one of the several talking machines that were named "Little Wonder," but had nothing to do with the popular Columbia-pressed records of the same name. The machine pictured on this mid-1910s flyer interestingly features a record clamp threaded to the turntable spindle: a feature dropped a decade earlier by the major companies. *Courtesy of Merle Sprinzen (Value code: K)*

4-39. On occasion, we are treated to a glimpse of a family's vacation accommodations from long ago. Talking machine advertising of the early twentieth century often suggested the benefits of prerecorded music during a summer idyll. This August 1915 image shows the porch of the bungalow at "Kamp Kareless" in Breckenridge, Minnesota, the summer home of Ransom Phelps. Judging from the pages of a journal kept at the time, the camp evidently drew family and friends like a magnet. Among the varied diversions was an aging Type "AH" Graphophone circa 1902, and a case full of records. See next illustration. *Courtesy of Paul W. Horgan (Value code: K)*

4-40. A nephew of Mr. Ransom Phelps kept a journal of his days at "Kamp Kareless" during the long-ago summer of 1915. We have him to thank for the concise yet vivid record of the family's vacation, and, evidently, the previous photograph. The final paragraph of the entry shown here seems particularly appealing (except perhaps the cleaning the fish). *...It was a beautiful moon light night on the lake. The music from the phonograph at the bungalow sounded fine on the water...* Courtesy of Paul W. Horgan

4-41. A very particular customer wanted his Victrola in a cabinet like no other. There surely is nothing about this Chinese chest with Queen Anne legs to suggest a talking machine. *(Value code: VR)*

4-42. Opening the doors and lid reveals the true nature of this artifact. Note the various drawers for sundries and record storage. The slats in the mouth of the horn seem inspired by Victor's own patented "Sound Amplifying Boards" which were first seen in the 1909 Victrola "XII," and incorporated into additional models during 1910.

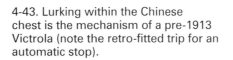

4-43. Lurking within the Chinese chest is the mechanism of a pre-1913 Victrola (note the retro-fitted trip for an automatic stop).

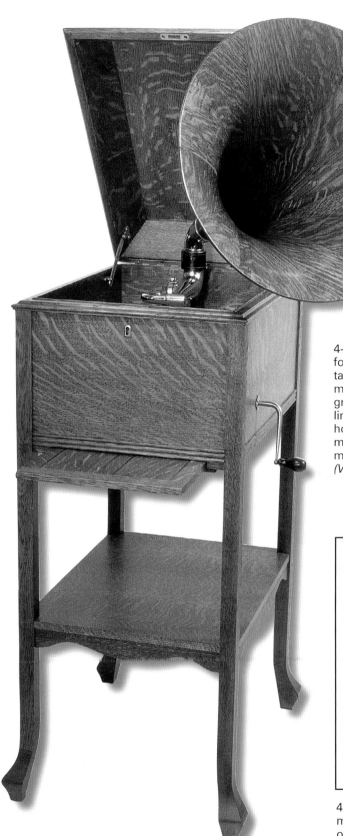

4-44. The Victor/Victrola "XXV" was introduced in August 1913 for $60.00. The instrument was promoted by the company as a talking machine for school or other educational use, and it was moderately successful in that role. Over the years, its cabinet grew in size, as did most others in Victor's extensive product line. This example is the earliest, smallest size. The smooth oak horn (as opposed to those with recessed features) is more commonly associated with the later, larger models. Closed measurements: 17" wide, 23" deep, 47" high. *Courtesy of Brice Paris (Value code: E)*

4-45. The short-lived alliance of Brunswick and Pathé Frères (a famous French company's American branch) is illustrated by the title of the record seen here with a 3 3/4" high metal Pathé promotional rooster. For a brief time in 1916/17, before Brunswick launched its own label in the United States, it distributed Pathé records, which resulted in this anomaly: "At Home With My Brunswick Phonograph" by Louis J. Winsch on the Pathé label! The same song, with "Pathé Pathéphone" substituted (and sung by Mr. Winsch), may be heard in an excerpt from an original recording featured on the CD accompanying *The Talking Machine, an Illustrated Compendium*, Fabrizio & Paul. *Courtesy of Don Fenske (Value code: K)*

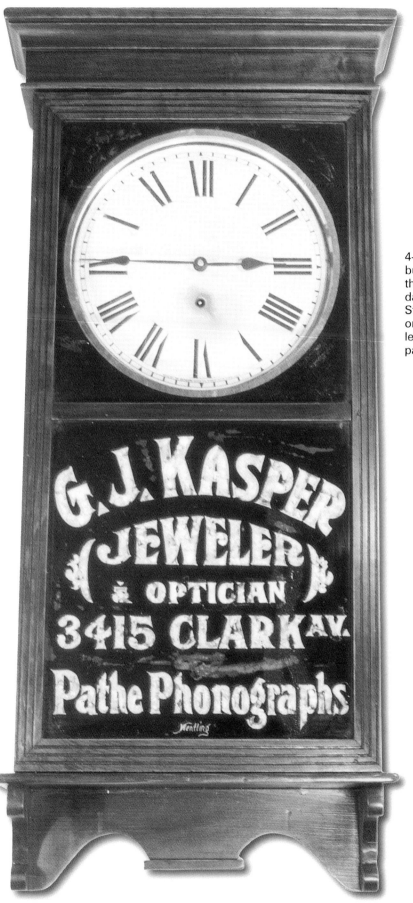

4-46. G. J. Kasper's Cleveland, Ohio, business must have been booming in the late-1910s; those were the halcyon days of the Pathéphone in the United States. This 17" x 38 1/2" clock was ornamented with faux-mother-of-pearl lettering, and inscribed by the sign-painter: "Wentling." *(Value code: VR)*

4-47. Souvenirs of encounters with famous recording artists can be ephemeral (autographed programmes, tickets or publicity photos), or enduring, if not monumental, such as the Electrola "XVI" seen here. *Courtesy of Jerry Blais (Value code, this particular, special example: VR)*

4-48. Opera star Luisa Tetrazzini was most likely making a personal appearance at a music store when she inscribed this Electrola with the same type of white ink customarily employed by celebrities to autograph recordings. A very large memento was thereby created. *Courtesy of Jerry Blais*

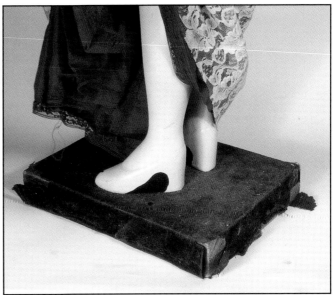

4-50. Ms. Galli-Curci stands atop a velour-covered base. Note the *trompe l'oeil* effect which suggested her high heels. *Courtesy of Don Fenske*

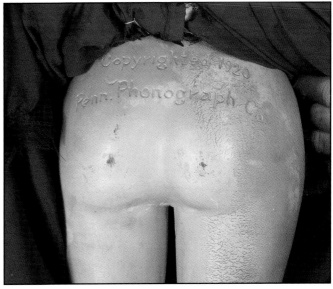

4-51. Common decency could not deter us from proving to any skeptics that this is not an ordinary store manikin. The Penn pedigree was prominently pressed into the prima donna's plump posterior. *Courtesy of Don Fenske*

4-49. Amelita Galli-Curci made her first Victor records in 1916, becoming a popular classical artist in the company's stable of luminaries. Lofty indeed was this 35" high plaster likeness of Victor recording artist Amelita Galli-Curci, formally attired right down to the understated pearls. *Courtesy of Don Fenske (Value code: VR)*

4-52. The second version of the Columbia "Grand" Grafonola cost $500.00 in mahogany, but was also available in expensive Circassian walnut for the more discriminating consumer. 42" long, 32" wide, 41" high. *Courtesy of Bob and Karen Johnson (Value code: VR)*

4-53. A number of furniture factories produced record cabinets based on a "sectional" or so-called "barrister" bookcase design. The Crescent Talking Machine Company, Incorporated, went one better and had a playing mechanism built into the top section. The undeniable convenience suffered a bit by the necessity of removing the crank before the upper section could be closed. The doors employed wooden panels rather than glass. 35" wide, 15 1/2" deep, 64" high. *Courtesy of Richard and Nancy-Ann Brown (Value code: VR)*

Bottom left:
4-54. The Crescent Talking Machine Company, disdaining the unmarked components of some of its rivals, had its soundboxes appropriately labeled. *Courtesy of Richard and NancyAnn Brown*

Bottom right:
4-55. Within the cabinet of the Crescent, the company name and trademark were proudly displayed. The Crescent Talking Machine Company was located at 89 Chambers Street, New York City. Advertisements first appeared in the *Talking Machine World* in August 1916. The trademark was not filed until August 15, 1919, but was in use since March 1914. *Courtesy of Richard and NancyAnn Brown*

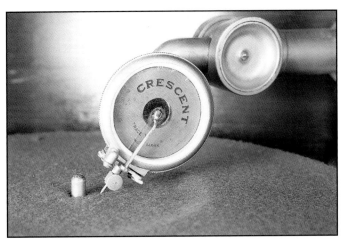

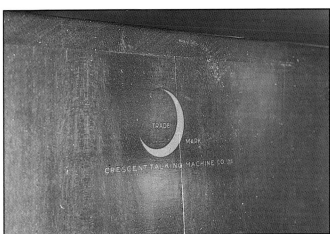

4-56. During the mid-to-late 1910s, the American public was showing interest in new styles of wicker furniture. At least one of the three biggest talking machine firms — Edison — had prototype wicker cabinets made for consideration, but no major American manufacturer put the potentially fragile cabinets into production. It was left to Heywood-Wakefield Company, already well established in the woven furniture trade, to market a line of phonographs in wicker cabinets. This interesting Heywood-Wakefield sits atop a matching record storage cabinet on casters. Note the unusually open weave of the lid's top. *Courtesy of the Johnson Victrola Museum (Value code: G)*

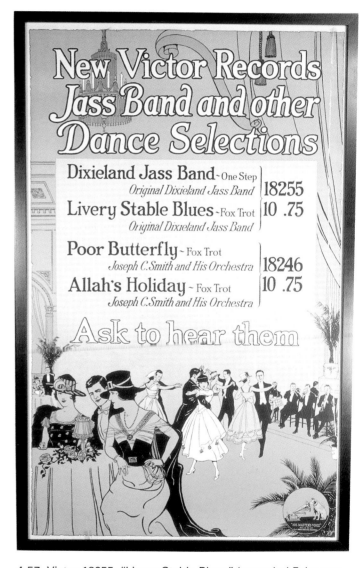

4-57. Victor 18255, "Livery Stable Blues" (recorded February 26, 1917), is regarded as the first jazz recording. Note that at this early date "jass" was the preferred spelling. The origin of this slang term has been debated, but is best described as a nonsense word with onomatopoeic power and rude sexual connotation. A poster, 20 1/2" x 33". *Courtesy of Daniel Melvin (Value code: J)*

4-58. In the heady late-1910s talking machine industry, many fledgling firms relied on a gimmick to gain the attention of potential customers. In this instance, the Crystola of Cincinnati, Ohio, rose above its generic mechanics and humdrum cabinet by offering something no other American manufacturer was supplying: an internal-horn made of glass. This is the "Model G." *Courtesy of Richard and NancyAnn Brown (Value code: G)*

4-59. With its grille removed, the shimmering beauty of the Crystola's mirrored horn is revealed. As with so many talking machine contrivances of the time, the difference in reproduced sound quality was dubious, but it was certainly eye-catching! Advertisements for the Crystola first appeared in the *Talking Machine World* in August 1917. *Courtesy of Richard and NancyAnn Brown*

4-60. A 6" high "Victrola" clock, with a Waterbury works. Inserting the images and effigies of talking machines into everyday contexts made good marketing sense. *Courtesy of Bob Carver (Value code: J)*

4-62. Disguised as a book, this European talking machine is neatly self-contained. The tone arm assembly stores to the left of the 5 1/2" diameter turntable, in an opening that is actually the mouth of the internal-horn. *Courtesy of Collection Fabrice Catinot, Dijon (Value code: I)*

4-61. This clock promoted the Beckwith-O'Neill Co., Minneapolis, Victor distributors. *Courtesy of Bob Carver*

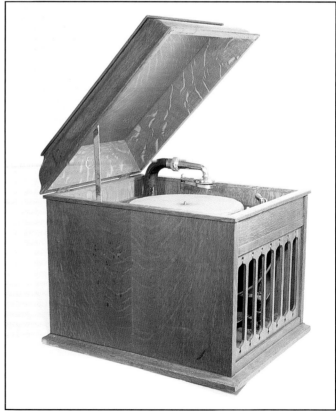

4-63. The Edison "Diamond Disc" "A-85" is not often encountered. The model is believed to have resulted from a February 1917 order for 500 special table Disc Phonographs for export (quartered oak, and wax finish). The "A-85" employed the "A-100" mechanism and its original price was $85.00. *Courtesy of Harvey P. Kravitz (Value code: VR)*

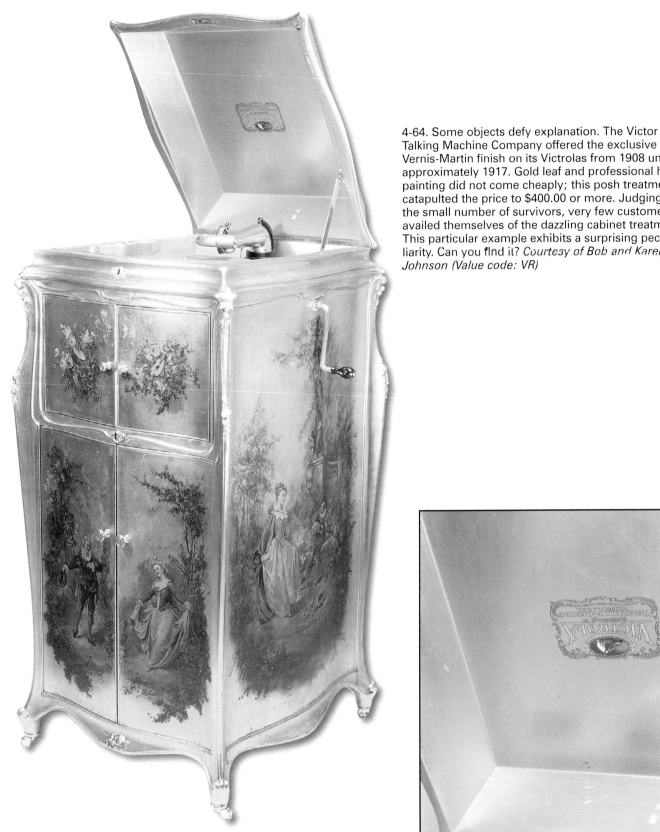

4-64. Some objects defy explanation. The Victor Talking Machine Company offered the exclusive Vernis-Martin finish on its Victrolas from 1908 until approximately 1917. Gold leaf and professional hand painting did not come cheaply; this posh treatment catapulted the price to $400.00 or more. Judging by the small number of survivors, very few customers availed themselves of the dazzling cabinet treatment. This particular example exhibits a surprising peculiarity. Can you find it? *Courtesy of Bob and Karen Johnson (Value code: VR)*

4-65. One must wonder what occurred at the factory when this Victrola "XVII" (No. 10958) was fitted with its lid decal. This is an anomaly we have not previously observed in the course of examining many hundreds of lesser Victrolas, and why this error was not remedied is a mystery — it would make more sense if it were found on a modest "VV IX," than on this luxury model. *Courtesy of Bob and Karen Johnson*

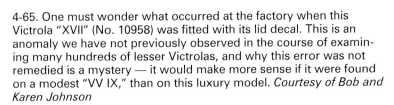

4-66. The production of Victrolas in oak was always a small fraction of the total. To emphasize the Victrola's elegance when it was introduced in 1906, mahogany was employed and remained the wood of choice. As the 1910s progressed, the several varieties of oak finishes became even less common in Victrolas, especially among the larger, more expensive models. It is truly rare to find a Victrola "XVII" in Golden Oak, as seen here. *Courtesy of Bob and Karen Johnson (Value code: VR)*

4-67. Custom Victrola cabinets are rare, but when encountered they usually conform to existing forms, with distinctive surface decoration. This Victrola "XVI" (No. 123131) is an exception. The corner columns were carved in oriental fashion, and the aprons similarly modified. The polychrome decorations are striking, and attest to the skill possessed by Victor's master decorators. Interestingly, the "Victrola" lid decal is absent. *(Value code: VR)*

4-68. Perhaps the most notable feature of this cabinet is the lid. Faux tiles and finials were added to suggest the roof of a pagoda. This additional ornamentation significantly added to the weight of the lid, and care must be taken while opening and closing. A little bit is known about its background — in the late 1930s, a Victor jobber in Indianapolis purchased this exceptional machine from its original owner, Mrs. Harold Onstott of Anderson, Indiana.

4-69. The reach of the Victor Talking Machine Company of Camden, New Jersey, was far-ranging. Often, its instruments and parts appeared in European incarnations, but sometimes ordinary United States models were sold in far-flung places, for instance South America. More exotic is this VV "IX" (circa 1920) which was sold in China, with accompanying Chinese-decorated record cabinet. The cabinet has been personalized with the name "J. Coz," suggesting it was a special order for the home of a European expatriate. *Courtesy of Phonogalerie, Paris (Value code, this particular example: VR)*

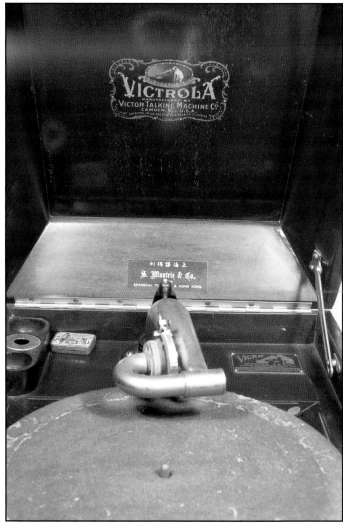

4-70. The interior of the VV "IX," with dealer label, "S. Moutrie & Co., Limited, Shanghai, Tientsin [modern Tianjin] & Hong Kong." *Courtesy of Phonogalerie, Paris*

4-71. The trade name "Operaphone" was employed in several capacities by talking machine firms. This elegant and unusual Operaphone instrument seems to be of European origin. Velour lined compartments under the lid accommodated 10" and 12" disc records. 33" long, 21" deep, 36" high. *Courtesy of Bob and Karen Johnson (Value code: F)*

4-72. The Shell-o-Phone Talking Machine Company of Chicago, Illinois must take the cake for a gimmick to sell phonographs during the post-World War One boom. The first advertisement for this instrument in the *Talking Machine World* appeared in October 1918. No gaudy cabinetry for these fellows, but rather a design conceived by Mother Nature. *Courtesy of Bob and Karen Johnson (Value code: VR)*

4-73. The decal illustrated not only the innovative amplifier employed by the Shell-o-Phone, but proclaimed the creation to be "The World's Greatest Musical Instrument." Heady stuff indeed, but something smells fishy… *Courtesy of Bob and Karen Johnson*

4-74. A potential customer for the Shell-o-Phone is asked to peek inside the cabinet as the salesman removes the grille, and...he must be kidding! For all its novelty, the genuine conch shell seems somewhat lost within the capacious cabinet, suggesting that perhaps an additional horn of wood was attached to direct the sound outside. Such a measure would have helped, but ultimately, the Shell-o-Phone was doomed to be a little fish in a big pond. *Courtesy of Bob and Karen Johnson*

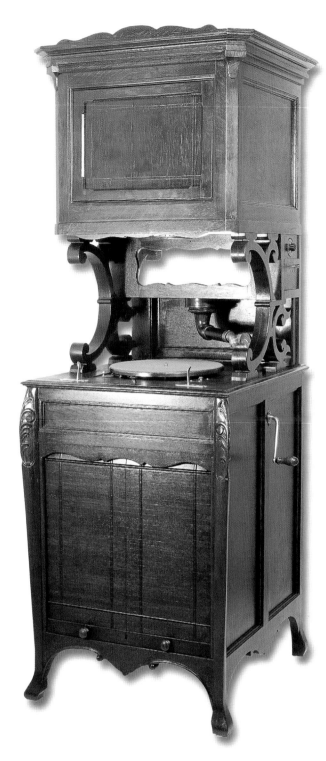

4-75. Just when we think we've seen it all… What appears to be a mirrored cupboard sitting atop a truncated, stylized Victrola is actually…? *Courtesy of Jon and Jackie Kelm (Value code: H)*

4-76. An Operatone! The lever to the left of the turntable opens the hatch covering the horn at top. The right lever is a volume control. The tambour door below lifts to access record storage. The designers of the Operatone clearly marched to the beat of a different drummer. Measures 22 1/2" wide, 25" deep, 64 1/4" high. *Courtesy of Jon and Jackie Kelm*

4-77. A close-up shows the switch at top right for the illuminating lamp above the turntable, and the locking accessory drawer. Note the needle compartment and cover in front of the turntable. Did we mention that the lower panel behind the machine contains a lockable storage compartment? This Operatone surely takes the award for imagination! *Courtesy of Jon and Jackie Kelm*

4-78. Of Sonora's distinctively curvaceous line of phonographs, the "Baby Grand" was the smallest, and therefore it is most unusual to see this model in Circassian walnut, a premium wood treatment. *Courtesy of Bob and Karen Johnson (Value code: VR)*

4-79. The Windsor Furniture Company, Chicago, offered a variety of striking phonograph cabinets equipped with generic mechanisms. This instrument is marked, "Pat'd Sept. 24, 1918." The first ad for this firm in the *Talking Machine World* appeared the following month. Measures 37 1/2" wide, 22" deep, 44 1/2" high. *Courtesy of Bill Landon (Value code: VR)*

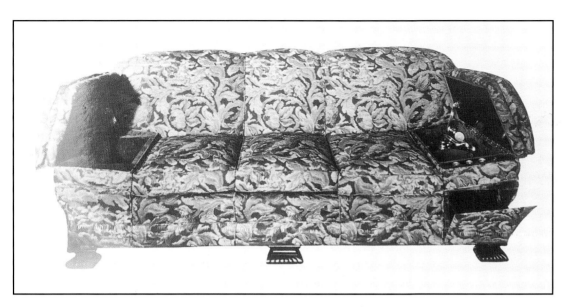

4-80. The Davis Upholstering & Furniture Company, located in Los Angeles, California, created what must be the most *comfortable* talking machine manufactured. An item in the September 1919 issue of the *Talking Machine World* illustrated and described the "…new and novel idea in home talking machines. Built into the arms of a large davenport, both the complete machine and a cabinet for records are conveniently accessible when wanted and at other times may be entirely hidden from view. The talking machine itself is complete in one end of the seat, the usual tone chamber [horn] being placed beneath the disc turntable and in the other arm is located a cabinet for the records. The tops of the arms, which are upholstered in the customary style, constitute hinged lids, which, when lowered, fit neatly over the mechanism and record box, respectively, and at the same time a cloth or tapestry-covered slide is provided for slipping back over the tone chamber [horn]. The davenport may, of course, be had in a variety of styles." *Courtesy of the Babcock House Museum and John H. Perschbacher (Value code: VR)*

4-81. This 18" x 26" dealer poster was indicative of the high-quality advertising Columbia was supplying during the late 1910s and early 1920s. The illustration is a copy of a pastel rendering done in an artistic manner. *Courtesy of Don Courtesy of Don Fenske (Value code: J)*

4-82. By the late 1910s, the talking machine had overtaken many other consumer goods in popularity and sales. This was recognized by the E.S. Applegate & Company when it marketed its household polish. The generic use of "Victrola" had found its way to the top of the label in large type. Below, the manufacturer recommended the polish for "Victrolas, Phonographs, Pianos, Furniture, Automobiles, Etc." clearly showing where it believed its largest customer base was to be found. At left, record lovers could make their records "look like new" again with the Everlasting Dustless Record Cleaner. *Courtesy of the Scott and Denise Corbett collection (Value code: K)*

4-83. By the late 1910s, several companies offered electric motors to run talking machines, or electric cranking devices for spring motors, all to eliminate the drudgery of manual winding. This Veeco (actually the Victor Electrical Equipment Company, Boston, Massachusetts — no relation to Camden) motor, complete with turntable, was designed to replace the spring-driven motor in a Victrola (models were available for Grafonolas as well). Shipping box measures 13 1/2" wide, 8" deep, 13" high. *Courtesy of Bill Landon (Value code: K)*

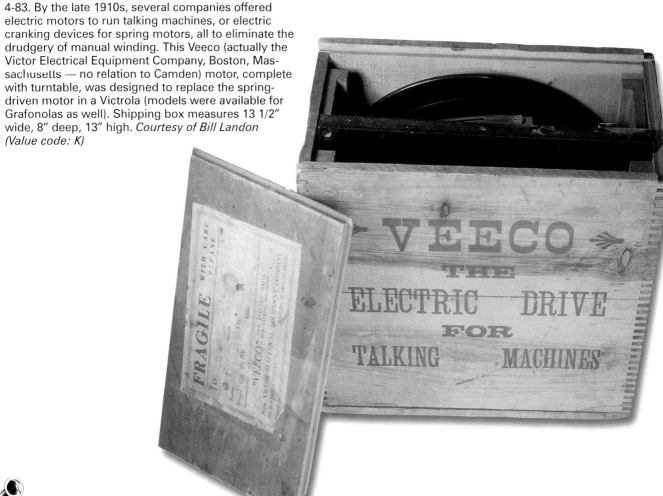

GOOD CONCERT MUSIC
For Mother and Dad when the Party is over

LULLABY (BRAHMS)
  Lashanska – 79114 – $1.00

POLISH DANCE
  Toscha Seidel – 78747 – $1.00

ROSES OF MEMORY
DREAMING ALONE IN THE TWILIGHT
  Maurel – A 2990 – $1.00

THAT NAUGHTY WALTZ
YOUR EYES HAVE TOLD ME SO
  Eddy Brown – A 2989 – $1.00

KISS ME AGAIN
  Rosa Ponselle – 49869 – $1.50

FAUST SELECTIONS
PARTS 1 AND 2
  Metropolitan Opera House Orchestra – A 6167 – $1.50

AIDA – O TERRA ADDIO (DUET)
  Ponselle – Hackett – 49734 – $2.00

Columbia Record

4-84. A Columbia counter display from Christmas, 1920. During the 1918-1921 period, Columbia's advertising was particularly colorful and attractive. 10 3/4″ x 14 3/4″. *Courtesy of Daniel Melvin (Value code: J)*

181

5-1. Sonora created a broad catalog of console phonographs in period furniture styles following the First World War. This Gothic model is distinguished by the amiable, carved heads that embellish the front doors. Although some of Sonora's attempts at fine furniture were lackluster and boring, this effort is, to our eyes, imaginative and appealing. *Courtesy of Bernard Clément (Value code: G)*

5-2. Pathé instruments were innovative and ingenious. The firm explored various methods for augmenting and directing the sound of acoustic talking machines. It employed reflected sound in its "Reflex" series, and paper cone amplification in the "Actuelle" and "Diffusor" models. This small "Diffusor" transfers the vertical cut vibrations of the sapphire stylus directly to the center of the cone, which slides laterally along a guide rod. The diameter of the cone is 14". In 1921, French patent No. 527886 was granted to Pathé for the "Diffusor" system. *Courtesy of Bob Carver (Value code: I)*

5-3. From November 1908 until the mid-1920s, Victrolas could be special ordered in Circassian walnut, a rare wood, at a considerable price premium. It's particularly unusual to find a console in Circassian walnut, such as this Model "230." The instrument carries a patent notice dated October 1, 1922, at which time interest in this deluxe treatment was on the wane. *Courtesy of Stan Stanford (Value code: VR)*

5-4. Open, the Circassian walnut Victrola "230" presents the eye with even more luxurious patterns. The serial number is 1000, a "round" figure we are tempted to think might have some special significance, but is more likely a coincidence. *Courtesy of Stan Stanford*

5-5. Victor's line of custom-made "Period" cabinet Victrolas was launched in 1917. Most were priced in the $300.00-$900.00 range, which put them out of reach of anyone but the wealthy. The company attempted to suggest certain "period furniture" styles ("Gothic" shown here), while maintaining a semblance of signature Victrola design components — with mixed results. Many "Period" cabinets ended up looking like a club sandwich of three Victrolas jammed together, with historical elements sprinkled on. *Courtesy of Jerry Blais (Value code: VR)*

5-6. The ID plates of the "Period" model Victrolas are worthy of note. The designations appear unconventional to those familiar with the plates of more ordinary Victor models. In what is thought to be standard company practice, Victor numbered its custom instruments beginning at an arbitrary "501," which would make this particular "Gothic" Victrola the thirty-second to have been manufactured. The redundant usage, VV-500, Gothic No. 500, is curious (see *Phonographs with Flair*, Fabrizio & Paul, figure 1-50). *Courtesy of Jerry Blais*

5-7. A lunchbox? A mailbox? No! A talking machine! This clever little Nirona from the 1920s snuggles into its handy 9" high carrier. *Courtesy of Garry James (Value code: I)*

5-8. Removed, it looks as if it would never fit back in. The crank may be seen in its upright storage position. The metal, grain-painted body is withdrawn using a handy ring. Note the tin of needles stored on the inside of the door. *Courtesy of Garry James*

5-9. The seldom-seen Edison Disc Phonograph "L-19" or "Louis XIV." Introduced in June 1919 at $300.00, this model was a slow seller, and a price increase to $350.00 in December 1919 didn't help matters. In March 1922 the company reduced the price to $295.00, but despite its walnut cabinet, No. 250 horn, and double-spring motor, the "L-19" elicited only lackluster sales. *Courtesy of Andrew S. Kida (Value code: VR)*

5-10. An Edison catalog of the period illustrated the "L-19" ("Louis XIV") with its slightly better-known sibling, the "W-19" ("William and Mary"). The company's marketing in the late 1910s and early 1920s stressed "Every Phonograph in a Period Cabinet." Although the "W-19" sold a bit better, had Edison engaged in effective market research, these two models might never have been built. *Courtesy of Gregg Cline*

5-11. Victor's $15.00 Victrola "IV" was the most portable of the line until 1921, when the Victrola "50" appeared. Until then, the traveling music lover either stuffed his Victrola in a trunk padded with clothing, or invested in a custom case such as this one by the Corley Company, Incorporated of Richmond, Virginia. *Courtesy of Don Courtesy of Don Fenske (Value code, VV IV in this particular case: VR)*

5-12. A close-up of the Corley trunk, showing method for securing the tone arm and "Exhibition" sound box. *Courtesy of Don Courtesy of Don Fenske*

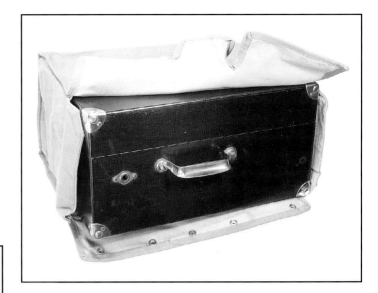

5-13. The Victrola "50" (introduced 1921) was constructed more substantially than most portable talking machines. It was available in oak or mahogany (shown) with a furniture-like finish. Yet, the fact that it was finely finished created the need for protection, since the instrument was intended to be taken out and about. To this end, note the custom-fit canvas sack enclosing this "50." *Courtesy of Bob Thomsen (Value code: J)*

5-14. Cylinder records in original packaging designed for the Averill phonograph mechanism used in the Madame Hendren, Mae Starr, and Dolly Rekord toy dolls of the 1920s. *Courtesy of the Tom and Sandi Mc-Carthy collection (Value code, each: K)*

5-15. A Mae Starr doll in original box. Note area on the label to indicate hair and eye color, and a section to fill in the name of its intended recipient. *Courtesy of Freddie and Jeanette Blair (Value code, average, unboxed doll: H)*

5-16. It's said that a picture is worth a thousand words. At first glance, there seems little to say, other than noting the tableau of a happy child with her toys, a small phonograph, her "Bubble Books" (full of kid's records), and her doll. Yet, a closer look reveals a rather sparsely decorated room; not the home of a "well-to-do" family. The child, however, is well dressed in a spotless shift and fresh bow. The photographer has positioned the camera not from the child's perspective, but *below* it. The man on his knees looking down into the reflex viewfinder of his camera was probably her father, regarding his daughter as though she were on a pedestal. Mother was likely standing behind, encouraging the little girl to smile. The talking machine and the "Bubble Books," though modest in price, may have been significant purchases for a family of limited means, but no matter — it was the child who was truly loved. *Courtesy of Merle Sprinzen (Value code: K)*

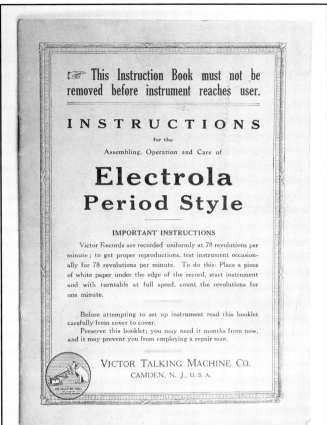

5-17. This instruction book accompanied a rare "Electrola Period Style" when it was shipped from the factory. One might wonder about the appearance of the machine for which this booklet was intended. For the answer, see the next illustration.

5-18. This photograph, taken in Victor's shipping department, shows a special-order Chinese "Period Cabinet" Electrola being crated for shipment. The booklet in the previous view was found with this very machine, which survives today in the following illustration.

5-19. As if by magic — a Chinese "Period Cabinet" Electrola in a crate, just as it was when it left the Victor factory — frozen in time!

5-20. This ungainly Victrola features a most un-Victrola-like cabinet. The label for the retailer — The Clark Music Company (a well-known Syracuse, New York, dealer) — implies that this machine was sold, not created in a home workshop — but what is it? *(Value code: VR)*

5-21. The answer is found by inspecting the data plate. The model is "SP-523" indicating a special order, possibly received in May 1923. True to Victor's serial numbering practice, the first machine off the line was No. 501. This may be the only surviving example of a Victrola "SP-523."

5-22. This behemoth was manufactured by George B. McConnel, Houston, Pennsylvania. The compartment to the left contains a generic motor and reproducing apparatus for 78 rpm records. The compartment to the right houses a post-1919 Edison "Diamond Disc" Phonograph mechanism. A built-in lamp provides plenty of illumination. A story accompanying this piece suggested it was used in a roller skating rink, but this massive piece of furniture also would have worked for a record retailer, or an early 1920s audio fanatic. Note the cranks on either side. Measures 55" wide, 24" deep, 52" high. *Courtesy of Marilyn and Robert LaBoda (Value code: VR)*

5-23. A Victor "Period" Victrola in the "Chinese Chippendale" style. Note that the decorations were not applied to a standard cabinet design, but that the entire creation was a custom rendering. This example is No. 502, indicating that it was the second of this model to be manufactured. 22" wide, 23" deep, 48" high. *Courtesy of Bob and Karen Johnson (Value code: VR)*

5-24. The hardware for the "Chinese Chippendale" "Period" Victrola was designed to harmonize with the rest of the instrument. *Courtesy of Bob and Karen Johnson*

5-25. A most unusual Aeolian-Vocalion mechanism in a custom American Walnut cabinet. The absence of a decal suggests that this striking cabinetry was not a product of the Aeolian Company. What appears to be a door knob on the front left is actually a volume control. 29" wide, 26" deep, 39 1/2" high. *Courtesy of Bob and Karen Johnson (Value code: VR)*

5-26. Silvertone was Sears, Roebuck & Company's house brand for phonographs. The line was meant to sell at low prices, so the moniker is not associated with high-end merchandise. Therefore, it's a surprise to discover this Silvertone with individually book-matched oak veneer doors, polychromatic touches, gilt highlights, and elaborate appliqués. Even the rich violet turntable velour, which Sears characteristically employed in its best models, suggests elegance. In the 1922 Sears catalog, a nearly identical model (with slight variation in grille and appliqués) was called the "Majestic" and sold for $175.00. It was described as being "Louis Fifteenth Period Art Design." 23" wide, 24" deep, 48" high. *Courtesy of Bob and Karen Johnson (Value code: G)*

EDISON REALISM TEST

1. State what kind of voice (soprano, tenor, etc.,) or kind of musical instrument you wish to hear.

2. Sit with your back toward the instrument.

3. Spend two minutes looking through the scrap book which will be handed to you by demonstrator.

4. Then select one of the clippings at random and read it carefully.

5. Having read the clipping, recall the last time you heard the kind of voice or instrument which you have asked to hear. Picture the scene. When it is clearly in your mind, say to the demonstrator, "I am ready."

6. About forty-five seconds after the music begins, close your eyes slowly and keep them closed for a minute or more. Then open your eyes for fifteen seconds but do not gaze at your surroundings. After this, close your eyes again and keep them closed until the end of the selection.

Result You should get the same emotional re-action experienced when you last heard the same kind of voice or instrument.

If you do not obtain this re-action at the first test, it is due to the fact that you have not wholly shaken off the influence of your surroundings. In that case you should repeat the test until you are no longer influenced by your surroundings.

5-27. In the late 1910s and early 1920s, Thomas A. Edison, Incorporated engaged in several pseudo-psychological marketing practices. One of these was the "Edison Realism Test," whose rather elaborate instructions are shown here. Once the customer had negotiated the timed stages of step six, his expected "emotional reaction" was described. If the unfortunate subject had not experienced the desired reaction, he was admonished to repeat the procedure until he got it right — "until you are no longer influenced by your surroundings." We wonder if the "surroundings" included the stopwatch, eager salesman, and scolding instructions! *Courtesy of the Jerry Koch collection*

5-28. Typical pages from the scrapbook mentioned in the "Edison Realism Test" instructions show Anna Case with the popular Edison "C-250"/"B-19" and the expected heavy-handed company propaganda disguised as newspaper clippings. *Courtesy of the Jerry Koch collection*

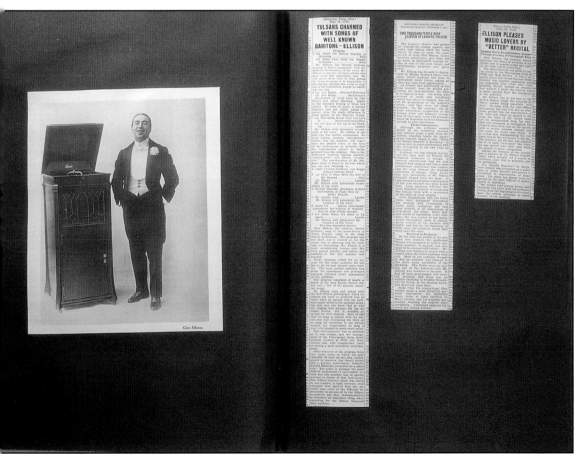

5-29. Same Phonograph, different artist, more persuasion! How much could a person endure before giving up and signing the purchase agreement — just to make it all stop? *Courtesy of the Jerry Koch collection*

5-30. From 1915 through the early 1920s, the flagship of the Edison "Diamond Disc" Phonograph line was the Model "C-250" (renamed the "C-19" in 1919). It was prominently featured in Edison advertising, "Tone Tests," and literature of the time. Here, a 4 1/16" high effigy of the popular model served as a bank. Banks in the shape of specific talking machines were used as sales tools — "save up to buy the instrument you've always wanted!" *Courtesy of George G. Glastris (Value code: J)*

5-31. In the early 1920s, the HomoPhone Company of Newark, New Jersey, offered a device whereby specially prepared wax discs could be recorded and reproduced at home. The dream of capturing on disc Aunt Matilda's warbling in the front parlor harkened back to the mid-1890s when the Berliner Gramophone Company had hinted of such things (though it was years before an effective amateur disc recording device would be designed). This HomoPhone in its carrying case still retains its original horn and instructions. *Courtesy of John T. Hoffman, the Phonophile (Value code: J)*

5-32. The illustrations of the HomoPhone's component parts show a device remarkably similar to the Bell-Tainter cylinder Graphophone of the late-1880s/early-1890s. As described here, the HomoPhone was designed to shave the wax discs as well as record them. *Courtesy of John T. Hoffman, the Phonophile*

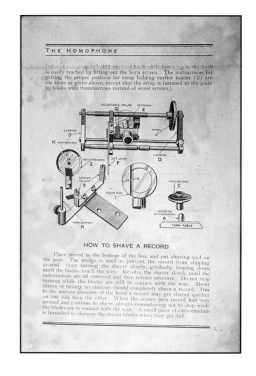

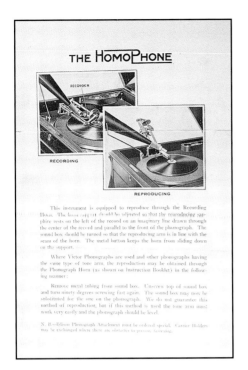

5-33. This page of the HomoPhone's instructions shows the device in place on the customer's Victrola. In addition to shaving and recording, the Homo-Phone could also reproduce, although the text warned, "We do not guarantee this method of reproduction…" In other words, it was recommended to use the regular soundbox and tone arm of the talking machine to play the recording. *Courtesy of John T. Hoffman, the Phonophile*

5-34. Beginning in the First World War period, the French "Concert Automatique" was a hugely successful style of coin-operated talking machine. These instruments were installed in cafes, hotels, and other public places. For ten centimes, the whole room got an earful of the latest music — the ostrich-like horn configuration greatly amplified the sound and directed it over the heads of the patrons. Automatic features were minimal — dropping a coin turned on the motor, a timer turned it off, and the patron manipulated the tone arm himself. The tambour front opened to reveal indexed record storage. Pathé made the machines for various companies which distributed them. *(Value code: E)*

5-35. For use in a movie or vaudeville house, a Pathé advertising lantern slide. The company aggressively promoted its niche — vertical cut reproduction. In truth, changing the steel needles required by lateral cut instruments was probably not as annoying as Pathé frequently suggested. Today, we are most struck by the high quality of Pathé recordings. *Courtesy of Daniel Melvin (Value code: K)*

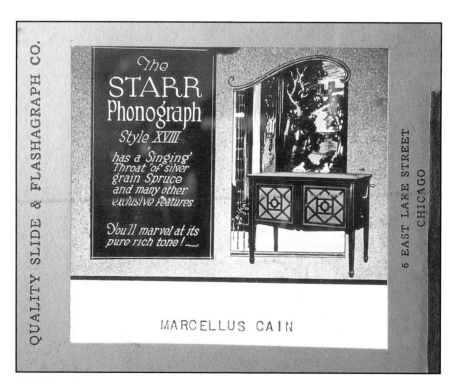

5-36. Manufacturers contended with discovering unique ways to describe competing instruments that were often similarly constructed. Starr's "singing" throat of silver grain spruce was poppycock. Any material could be used to fabricate an internal-horn. The most important factor affecting sound quality was the horn's mathematically conceived shape. *Courtesy of Daniel Melvin (Value code: K)*

5-37. French scientist Louis Lumière invented the pleated diaphragm reproducing system that bore his name. In England, in 1908 he patented, "A diaphragm for a phonograph consists of elastic material under torsional stress. The material is pleated and brought together to form a circular diaphragm with radial ridges and furrows." In 1912, he built an experimental model using a Gramophone Company, Limited instrument as a base. The First World War interrupted the project, after which an employee of the Gramophone Company, Limited, Mr. H.L.T. Buckle, contributed further to the pleated diaphragm reproducing system. British patent No. 187922 (Lumière & Buckle) envisioned a slanted motor mount, with the amplifying diaphragm mounted parallel to the turntable. Subsequently, however, all British models featuring the Lumière system were manufactured with the diaphragm in an upright position, at a right angle to the turntable (thanks to Lumière's 1923 patent No. 224856). Shown is the "His Master's Voice" Model "510," introduced in the fall of 1924. It was identical in appearance to Model "511" (with customary acoustic arm and soundbox) except that behind the front doors, in the space usually taken up by the internal-horn of a conventional talking machine, there was plenty of room for record storage. The Lumière models, falling as they did just prior to the introduction of the truly revolutionary "exponential" (in Great Britain, "re-entrant") horn in 1925, did not sell well and were withdrawn in 1927. *Courtesy of Jerry Blais (Value code: E)*

5-38. The Lumière instruments made in Great Britain and sold in France under the "La Voix de son Mâitre" (His Master's Voice) brand reverted to the original slanted motor/parallel diaphragm concept. Volume was cleverly controlled using the interior of the cabinet lid as a sound-reflecting dish. There were several friction stops on the support arm for the lid, which allowed it to occupy different positions, including fully up, and open a crack, thereby regulating the reflected volume. *(Value code: F)*

The following are Victor trolley cards from the late 1920s "Orthophonic era."

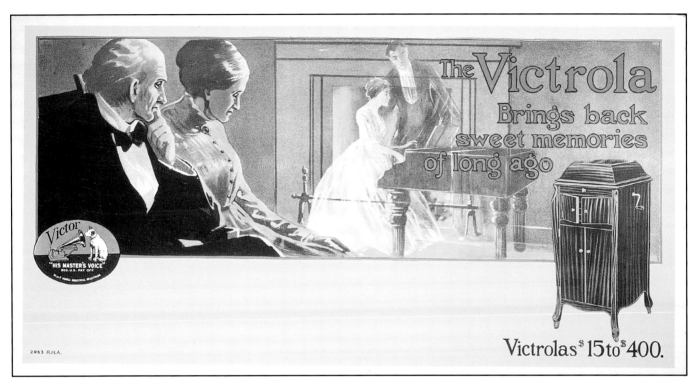

5-39. A head-on collision between the nineteenth and twentieth centuries: the Victrola! A "trolley card," measuring 10 1/8" x 20 1/2". *Courtesy of the Johnson Victrola Museum (Value code: I)*

5-40. *Courtesy of the Johnson Victrola Museum (Value code: I)*

5-41. *Courtesy of the Johnson Victrola Museum (Value code: I)*

5-42. *Courtesy of the Johnson Victrola Museum (Value code: I)*

5-43. *Courtesy of the Johnson Victrola Museum (Value code: I)*

5-44. *Courtesy of the Johnson Victrola Museum (Value code: I)*

5-45. When Victor inaugurated its new Orthophonic line in November 1925, the flagship
was, and would remain until 1928, the "Credenza," later known as the "8-30." Ordinarily,
this machine was priced at $275.00, but when
supplied with painted leather panels as is
this example, the cost was $325.00. Finally,
the electric motor option ($35.00) boosted
the total price of this "Credenza" to $360.00.
*Courtesy of Bob and Karen Johnson (Value
code: G)*

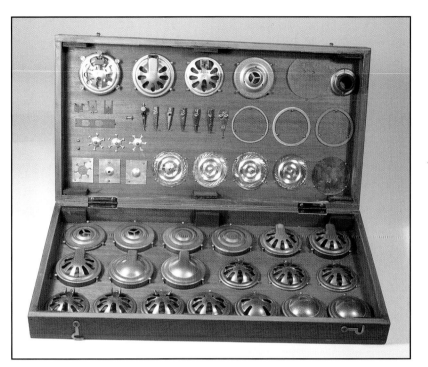

5-46. This interesting Victor Talking Machine Company display shows the stages of manufacture for the various components that comprised the "Orthophonic" sound box (introduced in November 1925). *Courtesy of the Johnson Victrola Museum (Value code: VR)*

5-47. What at first glance appears to be an oak gate-leg table is actually a very well-disguised talking machine. *Courtesy of the Domenic DiBernardo collection (Value code: VR)*

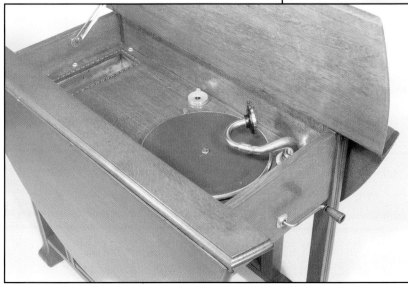

5-48. The center portion of the table lifts to reveal a mid/late 1920s playing mechanism and the horn opening at the far end. A celluloid plate displays (German?) patent No. 757546. *Courtesy of the Domenic DiBernardo collection*

5-49. Pathé kept up with the times, offering its version of the late-1920s cabinet talking machine. Victor had its "Orthophonic," Columbia had "Viva Tonal," Edison had "Edisonic," Brunswick had "Panatrope," and Pathé had "Olotonal." Measures 25" wide, 22" deep and 41 1/2" wide. *Courtesy of Phonogalerie, Paris (Value code: G)*

5-50. The German firm of Bing was a well-known manufacturer of toys. It also produced several toy talking machines during the 1920s, such as the well-known "Pygmyphone." Far more unusual is this charming "Little Dancer" with its antic horn. Overall height, 15". *Courtesy of Brice Paris (Value code: H)*

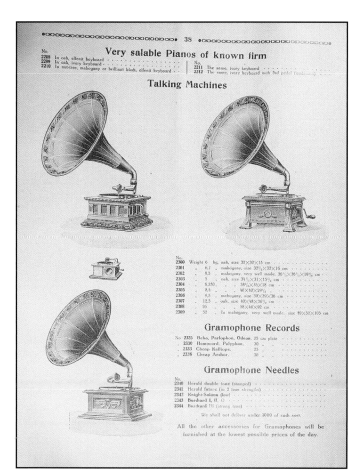

5-51. On April 12, 1927, the Levis Music Store, one of Rochester, New York's premier music merchandisers, received an "Illustrated Catalogue of Musical Instruments, Accessories and Strings of All Sorts" from L. Otto Reichel, Markneukirchen, Germany (whose company slogan was somewhat lost in translation: "From the good the best — at moderate prices"). Brass instruments, harmonicas, harmoniums and *external-horn talking machines*! Pictured on page 38 was the same type of Teutonic-looking apparatus that had been offered in the German market for the previous 20 years. By this point, Victor couldn't have cared about the taper arms clearly employed by these instruments; the patent wars for key acoustic talking machine features were over — and, except for a once-in-a-blue-moon order to L. Otto Reichel, the genre itself had all but vanished in the United States. *Courtesy of*

5-52. A British lantern slide advertising a Columbia portable, during the late-1920s portable phonograph "craze." *Courtesy of Daniel Melvin (Value code: K)*

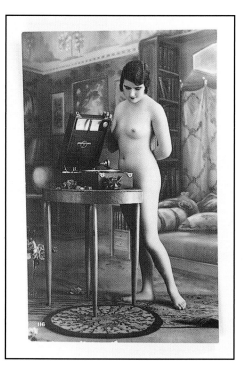

5-53. There are phonographs for music lovers of every stripe. Though this kind of pussy cat obviously has no stripes, who could doubt the sincerity of her affection for good music? *Courtesy of John T. Hoffman, the Phonophile (Value code: K)*

5-54. The field of portable phonographs in the late 1920s was rife with imaginative design and clever engineering. This "Majesta" featured a surprisingly long tone arm and a diminutive 6 3/8" diameter turntable. The combination of little ball feet, side-opening sound door, and cute brass hardware makes this portable particularly attractive. *Courtesy of Merle Sprinzen (Value code: J)*

5-55. The "Phono-Diff" operated like a mini-version of the Pathé "Actuelle," which was a type of Pathéphone related to but more complex than the "Diffusor" models. Many portables exploited reflex sound amplification because it was perfectly adapted to limited space, but at work here was a more complicated system, involving a resonating gold diaphragm. In the late 1920s a portable phonograph craze swept the world, but few designs took Art Deco to such heights. *Courtesy of Marie-Claude Stéger (Value code: I)*

5-56. From Deutsche Grammophon, circa 1929, an "Electrola" portable. Many collectors might find it hard to imagine how a portable could qualify for the nomenclature "Electrola" — which, in the United States, was reserved for Victor's expensive, electrically driven (and later electrically amplified) instruments. Yet, this modest-looking suitcase model embodies some sophisticated technology. It has *both* a spring motor and an electric motor (which can adjust to a considerable variety of mains power settings). Moreover, it could accept two styles of "needle head:" one was a conventional Victor/Gramophone Type "4a," that reproduced acoustically through a humble internal-horn (shown); one was an electric pickup for which an output jack was provided to play records through a radio. The radio provided the amplifier and speaker; a volume control was built into the portable record player. Sitting atop the turntable is a portable record "caddy," in which records could be stored for later selection. *Courtesy of Alan H. Mueller (Value code: I)*

5-57. Comrades, this is one stylish little talking machine! This revolutionary-red Russian record player was surprisingly compact, yet featured a swiveling cast metal horn. The red turntable is 6 3/8" in diameter. A Russian needle tin is to the right. *Courtesy of Merle Sprinzen (Value code: J)*

VALUE CODE KEY

Greetings, my friend. We are all interested in the future, for that is where you and I are going to spend the rest of our lives. And remember my friend, future events such as these will affect you in the future
— Criswell, from Ed Wood, Jr.'s film
"Plan 9 from Outer Space"

It's not getting any easier to attach values to antiques, nowadays. Having worked at it for 37 years, it's more daunting than ever. The antiques market has experienced revolutionary changes in the past ten years. For decades preceding the general popularity of the Internet, the market had been reasonably stable. Items of value tended to accrue worth in a predictable manner as the years passed. The Internet, and online auctions, although they brought with them great benefits, also have given us the "binge and purge" syndrome.

Let's say a certain cylinder record with the (imaginary) title of "My Red Hot Mamma Got Cold Feet at the Altar" has been coveted by collectors for years, but rarely found. One day it shows up on a popular online auction. The collecting community takes notice, and all marvel as, in the last seconds of the auction, "snipe" bids drive the selling price up to $3400. Collector chat rooms and newsgroups are ablaze with the exciting news that two stalwarts have battled it out to achieve an undreamed-of price for the record. In the "old days" this would have established a new benchmark for that particular record. In the brave *new* world, however, something entirely different can happen. Lots of people, collectors and merely interested parties, have noticed the result of the auction. This provokes some of them to pick through their boxes of old records, and... suddenly another copy of the formerly impossible-to-find cylinder appears in the online auction. This one brings only $1800 — still pretty high — but the episode is not over. Several more have apparently been discovered. By the time the cycle is complete, the value of this record has been utterly gutted. Binge and purge.

Fortunately, things like the above hypothetical example do not happen every day. Most items maintain their prices, and continue to grow in value. However, we point out the Faustian bargain we collectors have made with the Internet because it informs our daily lives. We draw much satisfaction from online auctions — they give us direct access to objects we never imagined we could own. Yet, the Internet has brought frustration, too — most collectors who use an online auction have some complaints about the way the system operates.

We have dealt with the job of creating a value guide as realistically as possible. Items which are extremely uncommon will be rated "very rare" (VR). The "VR" designation is not meant to indicate a price of any kind — it is only an expression of the infrequency with which the object changes hands. The particular theme of this book means that there will be a lot of "VR"s.

Values will be expressed *in the individual captions* by the following letter codes:

A   More than $15,000.
B $10,000. to $15,000.
C $6,000. to $10,000.
D $4,000. to $6,000.
E $3000. to $4,000.
F $2000. to $3000.
G $1250. to $2000.
H $750. to $1250.
I $350 to $750.
J $150. to $350.
K   under $150.

This might seem to be a long list of prices. However, you will find that many of the values actually will fall within a small range of categories.

By employing price ranges instead of specific dollar amounts, we leave ourselves open to various "slings and arrows" of discontent. It is our feeling that ours is the only effective approach to the subject. However, this does require the *discretion* of the reader.

In using the value codes, we advise *conservatism*. If you are trying to establish the value of an object in your possession, the natural instinct is to seize upon the highest price in the given range, regardless of condition. This can be a mistake. The condition of the items in the book is nearly always excellent, but the condition of *your* item is the important issue. Condition can affect the value by up to 50% — and occasionally more. The

bottom end of a price range has as much validity as the top, though it seems a less attractive prospect. Different examples of the same type of object may represent the entire range of a certain price category, due to condition. The ranges are also intended to be *comparative*. Certain objects in category "G," for instance, may *never* approach $2000.00, no matter how perfectly preserved they are.

Beyond condition, many, many factors affect the price of an "antique." Exact worth is something that cannot be fixed precisely in time. We can only hint at it. Our guide is intended to help you understand the value of the objects illustrated, but we vigorously discourage, "It sold for such-and-such a price on eBay." True worth is what the market will bear. Quoting books, auction results, or what a shop is asking will probably not elevate the price higher than the general market will currently support.

BIBLIOGRAPHY

Association for Recorded Sound Collections, various editions: *ARSC Journal*, various issues to present.

Barnum, Frederick O. III. *His Master's Voice in America.* Camden: General Electric Company, 1991.

Barr, Steven C. *The Almost Complete 78 RPM Record Dating Guide (II).* Huntington Beach: Yesterday Once Again, 1992.

Baumbach, Robert W. *Columbia Phonograph Companion, Vol. II (Disc Graphophones and the Grafonola.)* Woodland Hills: Stationery X-Press, 1996.

_____. *Look for the Dog.* Los Angeles: Mulholland Press, 2005.

_____. *The Victor Data Book.* Los Angeles: Mulholland Press, 2003.

Bayly, E., ed. *The EMI Collection.* Bournemouth (GB): Talking Machine Review, 1977.

_____. *The Talking Machine Review International.* Bournemouth (GB): various issues beginning 1970.

Bayly, Ernie, and Michael Kinnear. *The Zon-o-phone Record.* Heidelberg (Victoria, Australia): Michael Kinnear, 2001.

Bill, Edward Lyman, ed. *The Talking Machine World.* New York: various issues 1905-1929.

Brooks, Tim, ed. *Little Wonder Records, a History and Discography.* St. Johnsbury (VT): New Amberola Phonograph Company, 2000.

Bryan, Martin, ed. *The New Amberola Graphic.* St. Johnsbury (VT): various issues 1973 to 2003.

Caruana-Dingli, Mark. *The Berliner Gramophone, an Illustrated History.* Markham (Canada): published by Domenic DiBernardo, 2005

Catinot, Fabrice, and Claude Crenn. *Le Son... en Images.* Dijon (France): 2003.

Charosh, Paul, compiler. *Berliner Gramophone Records, American Issues 1892-1900.* Westport: Greenwood Press, 1995.

City of London Phonograph and Gramophone Society, various editors. *The Hillandale News* (beginning in 2002 *For the Record*). London: various issues 1960s to present.

Conot, Robert. *A Streak of Luck.* New York: Seaview Books, 1979.

Copeland, George A., and Ron Dethlefson. *Edison, Lambert Concert Records & Columbia Grand Records and Related Phonographs.* Los Angeles: Mulholland Press, 2004

Dethlefson, Ronald, ed. *Edison Blue Amberol Recordings, 1912-1914.* Brooklyn: APM Press, 1980.

_____. *Edison Blue Amberol Recordings, 1912-1914* (paper cover). Woodland Hills (CA): Stationery X-Press, 1997.

_____. *Edison Blue Amberol Recordings, 1915-1929.* Brooklyn: APM Press, 1981.

_____. *Edison Blue Amberol Recordings, 1915-1929*, second edition. Los Angeles: Mulholland Press, 1999.

Dethlefson, Ronald, and George Copeland. *Pathé Records and Phonographs in America, 1914-1922.* Los Angeles: Mulholland Press, 2000.

Edge, Ruth, and Leonard Petts. *The Collector's Guide to 'His Master's Voice' Nipper Souvenirs.* London: EMI Group, 1997.

Fabrizio, Timothy C., and George F. Paul. *Antique Phonograph Accessories and Contraptions.* Atglen (PA): Schiffer Publishing, Ltd., 2003.

_____. *Antique Phonograph Advertising, an Illustrated History.* Atglen (PA): Schiffer Publishing, Ltd., 2002

_____. *Antique Phonograph Gadgets, Gizmos and Gimmicks.* Atglen (PA): Schiffer Publishing, Ltd., 1999.

_____. *Discovering Antique Phonographs.* Atglen (PA): Schiffer Publishing, Ltd., 2000.

_____. *Phonographica: the Early History of Recorded Sound Observed.* Atglen (PA): Schiffer Publishing, Ltd., 2004.

_____. *Phonographs With Flair, a Century of Style In Sound Reproduction.* Atglen (PA): Schiffer Publishing, Ltd., 2001.

_____. *The Talking Machine, an Illustrated Compendium 1877-1929, Revised.* Atglen (PA): Schiffer Publishing, Ltd., 2005.

Fagan, Ted, compiler. *The Encyclopedic Discography of Victor Recordings, Pre-Matrix Series.* Westport: Greenwood Press, 1983.

_____. *The Encyclopedic Discography of Victor Recordings.* Westport: Greenwood Press, 1986.

Frow, George L. *Edison Cylinder Phonograph Companion.* Woodland Hills: Stationery X-Press, 1994.

_____. *The Edison Disc Phonographs and the Diamond Discs.* Sevenoaks, Kent (GB): George L. Frow, 1982.

_____. *The Edison Disc Phonographs and the Diamond Discs.* Los Angeles: Mulholland Press, paperback reprint 2001.

Gaisberg, Frederick W. *The Music Goes Round.* North Stratford (NH): Ayer Company Publishers, Inc., 1977 (1942 reprint).

Gracyk, Tim, with Frank Hoffmann. *Popular American Recording Pioneers, 1895-1925.* Binghamton (NY): Haworth Press, 2000.

Gracyk, Tim, ed. *Victrola and 78 Journal.* Roseville (CA): various issues 1994-1998.

Hatcher, Danny, ed. *Proceedings of the 1890 Convention of Local Phonograph Companies.* Nashville: Country Music Foundation Press, reprint, 1974.

Hazelcorn, Howard. *A Collector's Guide to the Columbia Spring-Wound Cylinder Graphophone, 1894-1910.* Brooklyn: APM Press, 1976.

_____. *Columbia Phonograph Companion, Volume I: Hazelcorn's Guide to the Columbia Cylinder Graphophone.* Los Angeles: Mulholland Press, 1999.

Hoffmann, Frank, and Howard Ferstler, eds. *Encyclopedia of Recorded Sound,* Second Edition. New York: Routledge, 2005.

Hunting, Russell, ed. *The Phonoscope.* New York: Phonoscope Publishing Company, various issues 1896-1900.

Johnson, E. R. Fenimore. *His Master's Voice Was Eldridge R. Johnson.* Milford: (DE), State Media, Inc., 1975.

Kelly, Alan, and Jacques Klöters. *His Master's Voice/De Stem Van Zijn Meester.* Westport (CT): Greenwood Press, 1997.

Koenigsberg, Allen, ed. *The Antique Phonograph Monthly.* Brooklyn: APM Press, various issues 1972-1993.

Koenigsberg, Allen. *Edison Cylinder Records, 1889-1912, 2nd ed.* Brooklyn: APM Press, 1988.

_____. *The Patent History of the Phonograph, 1877-1912.* Brooklyn: APM Press, 1991.

Laird, Ross. *Brunswick Records, a Discography of Recordings 1916-1931*, in four volumes. Westport: Greenwood Publishing Group, 2001.

Lorenz, Kenneth M. *Two-Minute... Cylinders of the Columbia Phonograph Company.* Wilmington (DE): Kastlemusick, Inc., 1981.

Marco, Guy A., and Frank Andrews, eds. *Encyclopedia of Recorded Sound in the United States.* New York: Garland Publishing, Inc., 1993.

Martland, Peter. *Since Records Began: EMI: the First Hundred Years*. Amadeus Press, 1997 (distributed by Timber Press, Portland, OR).

Marty, Daniel. *Histoire Illustrée du Phonographe (Illustrated History of the Phonograph).* Lausanne/Paris: Edita/Lazarus, 1979, (reprinted in various English language editions with slight modifications of the title).

Michigan Antique Phonograph Society, various editors. *In the Groove:* various issues to present.

Moore, Jerrold Northrop. *A Matter of Records.* New York: Taplinger Publishing Company, 1977.

Moore, Wendell, ed. *The Edison Phonograph Monthly,* various anthologies 1903-1916. New Albany (IN): Wendell Moore Publications.

Mould, Dr. Charles, ed. *The Galpin Society Journal.* St. Albans (UK): May 2005, "Augustus Stroh's Phonographic Violin, etc.," Alison Rabinovici.

Phono-Ciné Gazette, various issues 1905-1906 (reprinted by the Phonogalerie, Paris).

Proudfoot, Christopher. *Collecting Phonographs and Gramophones.* New York: Mayflower Books, 1980.

Reiss, Eric L. *The Compleat Talking Machine.* Chandler (AZ): Sonoran Publishing, (4th revised) 2003.

Rolfs, Joan & Robin. *Phonograph Dolls and Toys.* Los Angeles: Mulholland Press, 2004.

Rondeau, René. *Tinfoil Phonographs.* Corte Madera (CA): 2001.

Rondeau, René, ed. *The Sound Box.* Journal of the California Antique Phonograph Society, various issues to present.

Rust, Brian, and Tim Brooks. *Columbia Master Book Discography*, in four volumes. Westport: Greenwood Publishing Group, 1999.

Rust, Brian, compiler. *Discography of Historical Records on Cylinders and 78s.* Westport: Greenwood Press, 1979.

Sherman, Michael W., with William R. Moran and Kurt R. Nauck, III). *Collector's Guide to Victor Records.* Dallas: Monarch Record Enterprises, 1992.

Sherman, Michael W., and Kurt R. Nauck, III. *Note the Notes, An Illustrated History of the Columbia 78 rpm Record Label, 1901-1958.* New Orleans: Monarch Record Enterprises, 1998.

Sutton, Allan. *Directory of American Disc Record Brands and Manufacturers, 1891-1943.* Westport: Greenwood Press, 1994.

Sutton, Allan and Kurt Nauck. *American Record Labels and Companies, an Encyclopedia (1891-1943)* [with CD-ROM]. Denver: Mainspring Press, 2000.

Wile, Raymond R., and Ronald Dethlefson. *Edison Diamond Disc Recreations, Records and Artists 1910-1929.* Brooklyn: APM Press, 1985.

GLOSSARY

DISCOS MEJICANOS y ESPAÑOLES PATHÉ

AMBEROL: The first four-minute cylinder developed by Edison, 1908-1912. These cylinders are black in color, made of especially hard metallic soap, and were usually sold in green and white containers.

AMBEROLA: The name denoting Edison's line of internal-horn cylinder Phonographs, 1909-1929.

BACK MOUNT: A term used to describe a talking machine which uses a back bracket which supports the horn and thus removes its inertia from the SOUNDBOX. At the time, this configuration was generally known as "tapering arm."

BEDPLATE: The metal motor mounting plate (often a casting) to which the upper mechanism of a cylinder talking machine is mounted, and from which the motor is suspended.

BLANK: A smooth, grooveless unrecorded cylinder.

BLUE AMBEROL: The blue celluloid four-minute cylinder sold by Edison, 1912-1929. The containers for these cylinders were blue until 1917, orange and blue thereafter.

CARRIAGE: The assembly which "carries" the REPRODUCER of a cylinder talking machine across the recording.

CARRIER ARM: Edison nomenclature for the CARRIAGE. The HALF-NUT and its flat supporting spring are attached directly to the carrier arm on nearly all Edison machines.

COMBINATION ATTACHMENT: Devices offered by Edison to convert pre-1908 Phonographs to play four-minute cylinders in addition to the two-minute variety.

CONCERT: Edison's trade name for the five-inch diameter cylinder and the Edison Phonograph designed to play it.

CRANE: The support used to mount horns of 15-inch length or greater to cylinder talking machines.

CYLINDER: A geometric form on which entertainment recordings were made, 1877-1929. The earliest sheets of tinfoil were followed by self-supporting cylinders of ozocerite-covered cardboard, stearic acid/paraffin, hard metallic soap, and celluloid. Advantages of the cylinder format included a constant surface speed and the ability to make home recordings.

DIAMOND DISC: Edison's line of disc records and machines, 1912-1929. The Diamond Discs were vertically recorded, and the Diamond Disc Phonographs were one of the very few disc talking machines to use a FEEDSCREW.

DIAPHRAGM: The flexible, circular vibrating membrane of a SOUNDBOX or REPRODUCER which converts mechanical energy to acoustic energy or sound waves. Mica, glass, compressed paper, copper and aluminum were commonly used as diaphragms in early talking machines.

ELBOW: The connection between the horn and the SOUNDBOX or TONE ARM of a disc talking machine. Earliest elbows are made of leather, gradually giving way to metal elbows after 1900.

ENDGATE: A swinging arm which supports one end of the MANDREL on some cylinder talking machines. To mount or remove a cylinder it is therefore necessary to open and close the endgate.

FEEDSCREW: A threaded rod which usually drives a half-nut fixed to the carriage of a cylinder talking machine, thus guiding the reproducer across the record grooves. Certain disc machines used FEEDSCREWS to drive the SOUNDBOX across the record or to move the turntable beneath a fixed soundbox. Similarly, some cylinder machines used feedscrews to drive a mandrel longitudinally beneath a fixed reproducer.

FRONT MOUNT: A term used to describe a disc machine where the horn attaches directly to the SOUNDBOX, and the support arm (or mount) runs directly below and parallel to the HORN. In such an arrangement the support arm points in the same general direction as the bell of the horn. At the turn of the twentieth century, this configuration was known as "straight arm."

GRAMOPHONE: Originally the name by which Emile Berliner's disc talking machine was known, it became a generic term to denote any disc playing talking machine, but fell out of use in the United States.

GRAND: Columbia's trade name for the five-inch diameter cylinder and the various Columbia machines designed to play it.

GOVERNOR: The mechanical assembly in a talking machine motor which regulates the speed, usually by limiting the outward movement of spinning weights.

HALF-NUT: An internally threaded metal piece usually in the form of a nut which has been cut in half. The threads of the half-nut correspond to the threads of the FEEDSCREW. As the feedscrew revolves, the half-nut will be propelled along it, thus driving the CARRIAGE of a talking machine.

LATERAL RECORDING: The type of disc talking machine recording where the information is encoded in sides of the groove. (see VERTICAL RECORDING)

MANDREL: The tapered drum upon which the cylinder record is placed for playing.

NEEDLE: The point (usually steel) of the SOUNDBOX which rides the grooves of the recording and transmits vibrations to the DIAPHRAGM.

NEEDLE HEAD: A colloquial term for the SOUNDBOX or PICK-UP which accepts vibrations directly from the record.

NEEDLE TIN: A container to store and dispense steel talking machine needles.

NEEDLE SHARPENER (OR CUTTER): A device for re-pointing disc talking machine needles of fiber, thorn or steel.

PHONOGRAPH: Originally, the name given by Edison to the sound recording/reproducing device he invented in 1877. In American usage, the term has been applied to sound reproducing equipment in general, regardless of age or type of record played.

RECORDER: A device comprising a DIAPHRAGM and cutter stylus for recording cylinder or disc records.

REPEATING ATTACHMENT: A device for cylinder or disc talking machines which returns the NEEDLE or STYLUS to the beginning of the record after the selection has played.

REPRODUCER: The component comprising STYLUS, linkage and DIAPHRAGM, which reproduces the sound from the record grooves. This term is most frequently applied to cylinder machines. (See SOUNDBOX)

SHAVER: On a cylinder talking machine, a device which holds a knife in position to pare the recorded surface from a wax cylinder.

SOUNDBOX: The component comprising NEEDLE, linkage and DIAPHRAGM, which reproduces the sound from the record grooves. This term is most frequently applied to disc machines. (See REPRODUCER)

STYLUS: The point (usually sapphire or diamond) of the REPRODUCER which rides the grooves of the recording and transmits vibrations to the DIAPHRAGM or PICK-UP.

TALKING MACHINE: A general term for sound reproducing devices playing either cylinder or disc records, made prior to the advent of electric reproduction. Specific brands such as Graphophone, Victrola or Amberola all fall within the general category of talking machines. Modern American usage substitutes PHONOGRAPH for the older term "talking machine." For the purposes of historical accuracy, both terms will appear frequently in this book.

TONE ARM: A movable hollow tube which conducts sound to the HORN from the SOUNDBOX.

TRUNNION: Rightly, the sleeve visible on either side of the CARRIAGE which slides over the FEEDSCREW shaft on a Graphophone, but commonly used to denote the entire carriage.

TURNTABLE: The round platter which supports a disc record and revolves while the record is being played. After 1960, that separate component in a SOUND SYSTEM which comprised the TURNTABLE and RECORD CHANGER was often referred to simply as the TURNTABLE.

VERTICAL RECORDING: The type of cylinder or disc recording in which the sound vibrations are encoded in the bottom of the groove. (see LATERAL RECORDING)

WAX: Although the material from which many cylinder records were made is generally referred to as "wax," it was more rightly various formulations of metallic soap. The resemblance of this material to conventional wax promotes the common descriptive term.

INDEX

5-58. Horace Sheble (pronounced "sheb-lee") had a long, and occasionally arduous, career in the talking machine business. He began selling Edison Phonographs in Philadelphia in the late-1890s, in collaboration with partner Ellsworth Hawthorne. Hawthorne & Sheble became a huge force in the talking machine industry, maintaining several manufacturing facilities that produced a wide range of horns, cabinets, accessories, talking machines and records. One of the plants was in Bridgeport, Connecticut, home of the American Graphophone Company, a chief rival. The H&S enterprise came to grief in 1909-1910, a victim of the terrible swift sword of Victor's patent-litigation attorneys. Hawthorne went on to sell accessories for bicycles and motorcars. Sheble joined the camp of the Graphophone Company, and functioned as plant manager until he retired in 1912. This 5" diameter recording (both sides are "Auld Lang Syne") was given to the worthies assembled at Sheble's farewell dinner. (Value code: VR)

5-59. The menu, printed on the reverse label, includes fare chosen for its association with the State of Connecticut and , for old time's sake, the Philadelphia area. The Clover Club cocktail originated at the Bellevue Hotel bar in the City of Brotherly Love – an epithet called to mind by the menu's "Quaker Cured potatoes." Nearby Norristown and Mt. Airy are also referenced, and so forth. We can imagine that for Mr. Sheble it must have been a relief to have completed his life's work in his chosen field of enterprise, and to be looking forward to less demanding pursuits. And so we say, "Good-bye."

GRAMOPHO
ET
ACCESSOIRES
EDISON
BLUE AMBERO
RECORDS
LA VOIX DE SON MAITRE
COMPAGNE FRANÇAISE DU GRAMOPHONE
15 RUE BLEUE
PARIS
Catalogue Appareils
GRAND PRIX
Expon Univlle PARIS 1900
PHONOGRAPHES PATHÉ
98 RUE RICHELIEU PARIS
alk-o-Phone
LEARNING SOME NEW ONES.......
hone Company : Toledo, Ohio
Compa Francesa del Gramophone
Barcelona
Aparatos y Accesorios
Este Catálogo anula los anteriores
EDISON'S
PHONOGRA
All Talking Machines.
Edison-Bell Consolidated Phonograph Co., Ltd.,
39, CHARING CROSS ROAD, LONDON, W.C.